Oil, Jihad and Destiny

"This is a pungent book that does not mince its words, and rightly so. It is in fact an urgent wake-up call which we ignore at our peril. It is essential reading for the widest audience because what unfolds will affect everyone from the Cabinet Minister to the Kindergarten teacher; from the Preacher to the Policeman. Above all, it should find a place in the classroom because it is the new generation that faces the transition head on and deserves some guidance. The public desperately needs to be properly informed, first to give their governments the mandate for tough new measures, and second to avoid the risk of over-reaction....".

Colin Campbell
Founder of the Association for the Study of Peak Oil (ASPO)

Oil is the Achilles heel of western civilization. The world economy will collapse without great quantities of this black liquid. Although oil extraction and refining operations may be unpleasant, we must remember that oil fueled the Green Revolution that has fed millions of people on our planet. Access to oil frequently determines the well being, national security, and international power for those who possess it. Although the spectacular performance of the U.S. economy is the result of free enterprise optimism, it has been powered by oil, - billions of barrels of oil. As detailed in this book, the consequences of an energy shortage in the United States are draconian and any efforts to lessen its impact are essential.

Joseph P. Riva
Former advisor on world oil and gas
for the Congressional Research Service

What is this book about?

If we were to list the most important issues facing humanity, oil depletion has to be in the top three. The economic and cultural destiny of mankind is inexorably tied to the availability of oil. It is impossible to address the problems of famine without petroleum for fertilizer, cultivation, and processing. Oil provides the feedstock for thousands of products, including cosmetics, medicines, plastics, heating and cooking fuels, and mobile fuels for transportation.

But the days of surplus oil are coming to an end.

World oil is transitioning from a market driven by consumer demand to one limited by producer capacity. As a result, oil exporting countries are now able to control the price and the availability of an increasingly scarce commodity. Corporate behavior, government action, cultural stability, economics, legal agreements, geography, weather, crude oil transportation, military diplomacy and the always potent combination of religion and politics are now more important than geology in developing oil production forecasts.

The approaching oil crisis will have a global reach, impacting the economic and cultural health of every nation. Because they have the most energy intensive economies, however, the industrial nations of North America, Europe and the Pacific Rim will suffer the greatest deterioration.

We can either try to manage a "soft landing" or let nature take its course. Doing something means encouraging new attitudes about fuel production and consumption as well as the privilege of parenthood on a worldwide basis. If we do nothing, chronic recession is probable. Economic depression is possible.

This report provides a comprehensive examination of oil reserves and production, reviews the cultural challenges of the Middle East, analyzes the economic impact of four alternative oil depletion scenarios, and outlines a proposed course of action to enable a "soft landing". World oil production and consumption are evaluated by geographic region. This evaluation, along with a projection of how oil depletion could influence inflation, unemployment, economic growth and the price of gas, is presented in 8 tables and 32 charts.

Ronald R. Cooke

Oil, Jihad and Destiny

Will declining oil production plunge our planet into a Depression?

Ronald R. Cooke

Copyright and Credits

Published by
Opportunity Analysis
e-mail analysis@wizwire.com

Cover design by Iconix: www.iconix.biz

TABLE OF CONTENTS

CHAPTER 5   ISLAMIST CHALLENGE............................ 67

CHAPTER 6   HIGH PROBABILITY 80

Acknowledgements

To my wife, Cynthia, who would not let me give up.

To Colin Campbell, with whom I exchanged multiple e-mails and had two lively discussions. He helped me to focus on the subject at hand and offered several points which I have incorporated into the text of this report.

To Walter Youngquist, a man who is truly concerned about the future of Homo Sapiens and encouraged me to continue the crusade he started several decades ago.

To Seppo Korpela, the author of a presentation on oil depletion which was made to the Ohio Petroleum Marketers Association at their annual meeting in Columbus, Ohio in 2002, and a thoughtful contributor to this report.

To Joseph Riva. I found his excellent report for the Congressional Research Service, "World Oil Production After Year 2000: Business As Usual or Crises?" long after I completed my research and analysis for this report. It was a relief that someone else had independently reached essentially the same conclusions.

To Glenn Morton, with whom I had several exchanges, for his help in guiding my initial thoughts on this subject.

And to multiple friends who read and reread my literary efforts, always offering encouragement even though they were troubled by its essential themes.

Foreword

This is an important book in many respects. Foremost is that it draws attention in no uncertain terms to the devastating impact of oil depletion. Oil came onto the scene 150 years ago when the first wells were drilled. Production grew at an accelerating pace as the oil lands of the world were opened up, driving almost all aspects of the modern world with its unquenchable thirst for cheap and convenient energy. But now we face the second half of the Oil Age as production starts its inevitable decline towards eventual exhaustion at some point in the distant future. It is a polemic book with a powerful message, and rightly so, being written by someone trained in market research, who seeks to identify all the factors involved.

The second major contribution is to draw attention to the extremely unreliable nature of public data on the subject. The oil companies have under-stated discovery to comply with strict Stock Exchange rules, while some OPEC countries have exaggerated as they vied with each other for quota. They may even be reporting what they have found, not what remains, which would explain why the reserve numbers have barely changed since the implausible jumps of the late 1980s. Given the central role of oil in the modern world, it is truly a shock to find that the reserve book-keeping is so grossly unreliable, denying government planners even a minimal insight into the intractable problems for which they are responsible. With better information, the nature and impact of depletion would be self-evident. It would be easy to determine how much time we have within which to adjust to declining supply.

But things are changing. For example, ExxonMobil has published that discovery reached a peak in the 1960s to be followed by a relentless decline ever since. Extrapolating this firm trend gives a good indication of what is left to find and produce. Oil has to be found before it can be produced, so the peak of discovery forty years ago delivers a strong message that the corresponding peak of production cannot be far away.

The book presents a series of scenarios. Readers will want to review them in relation to the validity of the input assumptions. Scenarios are not forecasts but logical constructions, and their main purpose is to stimulate new interest and analysis. More reliable information will inevitably impact the results. If the resource base turns out to be lower than assumed, the peak and onset of decline will be that much sooner.

The importance of the Middle East is rightly stressed, with five countries bordering the Persian Gulf holding a large share of what oil is left, which puts their wealth of reserves in marked contrast with heavily depleted North America. Chapter 5 addresses this conflict by taking what can be described as a "homeland" view of evil Muslim extremists, backed by Saudi Arabia, who threaten the world's oil supply. Others may see the attack on Iraq as the culmination of a long-planned strategy to support of Israel in its conflict with indigenous Palestinians.

In either case, the region is a tinder box where mistaken policies could have quite devastating consequences on oil supply and its price, and indirectly the state of the world economy.

The book ends with some valuable recommendations, proposing an international agreement for the proper management of depletion with fair shares of the remaining supplies for all. It rightly points out that there are two sides to the equation. The producers for their part should endeavor to make the supplies last as long as possible and ease the tensions associated with the onset of decline. Consumers have an even more important role to cut waste, find new ways to use less oil, and encourage the development of alternative sources of energy.

It is no mean challenge to change direction and apply new mindsets in order to reject the excess and profligacy of the past. The open market was supremely successful in driving growth, but it has yet to develop an equitable mechanism by which to manage a decline imposed by dwindling energy supplies as imposed by Nature. The global market follows the precepts of classical economic theory dictating that the resources of any country should be accessible to the highest bidder, but as natural depletion begins to bite in earnest, communities will have find more self-sufficient solutions for energy.

This is a pungent book that does not mince its words, and rightly so. It is in fact an urgent wake-up call which we ignore at our peril. It is essential reading for the widest audience because what unfolds will affect everyone from the Cabinet Minister to the Kindergarten teacher; from the Preacher to the Policeman. Above all, it should find a place in the classroom because it is the new generation that faces the transition head on and deserves some guidance. The public desperately needs to be properly informed, first to give their governments the mandate for tough new measures, and second to avoid the risk of over-reaction. It would compound our problems if they misread the immutable laws of physics and the finite nature of the Planet that are driving us inexorably toward depletion. As to the Islamist threat. Oil depletion is happening with, or without, it. Middle East Muslim life is as foreign to Americans as the American way of life is to Middle East Muslims. Perhaps that explains why the parties have little in common and mistrust each other.

C. J. Campbell
Retired oil geologist and executive (FINA)
Founder of the Association for the Study of Peak Oil (ASPO)
County Cork, Ireland
April 2004

Prologue

Crude oil is created by nature at great depths below the surface of the earth from organic material. It then slowly migrates to the surface resulting in oil seepages, tar pits, and tar lakes. At the surface it is oxidized and eventually broken down, a process that provides nutrients to a variety of organisms. Only in rare circumstances is the migrating oil trapped and thus preserved for eventual exploitation by man. Less than five percent of the world's oil fields originally contained about 95 percent of total known world oil. These giant fields, because of their anomalous geology, were usually discovered early in an exploration cycle. They rapidly provided enormous amounts of oil from a relatively small number of wells. Currently, 20 percent of the world's oil supply comes from 14 fields that are, on average, almost 60 years old.

Oil is the Achilles heel of western civilization. The world economy will collapse without great quantities of this black liquid. It fuels transportation and industry, international politics, national treasuries, and military operations. Although oil extraction and refining operations may be unpleasant, we must remember that oil fueled the Green Revolution that has fed millions of people on our planet. Jet fuel, diesel fuel, pesticides, herbicides, fertilizer, plastics, and a huge number of other essential products such as medicines, paints, and clothing are made from petroleum products. Access to oil frequently determines the well being, national security, and international power for those who possess it. Although the spectacular performance of the U.S. economy is the result of free enterprise optimism, it has been powered by oil, - billions of barrels of oil. As detailed in this book, the consequences of an energy shortage in the United States are draconian and any efforts to lessen its impact are essential.

Oil will continue to have a vital role in the fate of nations. Since much of the world's remaining oil is concentrated in the Middle East, this area has become a strategic planning focal point for most of the industrialized nations. Saudi Arabia must play a leading role in future production.

> The entire world assumes that Saudi Arabia
> can continue to supply everyone with cheap energy.
> If this turns out not to be the case,
> there is no "Plan B".

There is no other standby oil production capacity. However, Saudi Arabia's five super giant oil fields (which to now have produced 90 percent of Saudi oil) were found between 1940 and 1965. Water injection into

Ghawar, the biggest field in Saudi Arabia, was 20% of the fluid pumped from this field in 1990. The water content has since increased to 60%, indicating the pending exhaustion of readily available oil. Although exploration beyond these five fields has intensified, only a handful of new fields have been found.

Over half of the oil consumed in the United States is imported. However, in spite of its dependency on a petroleum based economy, the United States is the only country in the world that restricts the search for oil. There is vigorous opposition to most Alaskan and offshore petroleum operations. However, big energy projects are not the greatest threat to the environment. The greatest threat, as detailed in this book, is the widespread poverty which a lengthy energy shortage would produce.

Energy does not as yet have a constituency, but that will change when the lights go out.

Joseph P. Riva, a retired petroleum geologist, and was an advisor on world oil and gas for the Congressional Research Service of the Library of Congress. He has testified before Congress and authored over 200 publications. He served on the Committee On Undiscovered Oil and Gas Resources, of the National Academy of Sciences, and the Interagency Coordinating Committee of the World Energy Program of the U.S. Geological Survey.

Author's Notes

Trademarks and Copyrights - This report contains material which belongs to the sources found in the text, footnotes, and References section of this report. The product and corporate names used herein, including trade marks, sales marks and source material copyrights, are the property of their respective owners. Every effort has been made to give proper attribution.

Purpose of this study- The objective of this research effort was to characterize the size and direction of the worldwide market for crude oil, including the depletion of reserves and the impact that alternative depletion scenarios would have on oil production and consumption. These scenarios were then analyzed in order to determine their impact on world and national economies, with a special focus on the United States.

Methodology - Market or industry research is a process that involves a number of interrelated steps. Research reports are based on facts and opinions which have been compiled, organized, analyzed and interpreted by someone who understands the research process. Market and Industry Research is not about absolutes. It's about people. Events. Products. Issues. Questions. And their interrelationship. It has FOCUS. It will deal with a specific question or issue.

In this study, it quickly became evident there were three key issues. How much oil can we expect to produce over the next 20 years? Will Islamic cultural conflict disrupt oil production? What is the economic impact of declining oil production?

This report is almost entirely based on fourteen months of secondary research. This research included a review of two books by Robert Baer, a book by Jan Goodwin, and selected reading from several books and reports on Arab history and culture. My research efforts also included information published by oil industry participants, trade journals and newspapers, various Internet sites, and data published by oil industry consulting firms. My primary research included several person to person interviews and topical discussions with industry participants.

I took a bottoms up analysis of the collected data. Of primary interest was the interaction of oil production with oil consumption and the subsequent impact that disruption and depletion would have on selected economic conditions. In order to accomplish this task, I developed a complex model of worldwide oil consumption and production that helped me to characterize alternative depletion scenarios. In order to verify long term trends, the model includes data that - in some cases - goes back to 1970. Economic impact columns include Gross Domestic Product (GDP), unemployment, inflation and gasoline prices for the United States, as well as world prices per barrel of oil. A more complete description of the interactive oil model can be found in Chapter 3.

In this report we deal with oil in its liquid form (crude oil) and in its highly viscous form (as it is found in tar sands and bitumen).

Market Segmentation- The data for this study was organized by four producer regions - North America, Middle East, EurAsia and Rest of World, and by four consumer regions - North America, Western Europe, Asia Pacific and Rest of World. Each region was analyzed as though it were an independent player in the market for crude oil. A definition of the primary nations that belong within each of these regions may be found in the "Definitions" section of this report.

Assumptions- The marginal quality of available data has forced me to make a number of educated assumptions as I developed the material for this report. These assumptions are presented and qualified as appropriate.

Disclaimer- There are multiple and sometimes conflicting opinions about the governments, religions, cultures, organizations, companies, markets and products discussed in this report. Specific or inferred references to, and all discussion of, persons, groups, organizations, events or circumstances are subject to a variety of interpretations and should therefore be treaded as conjecture. The reader should pay careful attention to the definitions described in the Appendix when interpreting the information contained herein. As of this date, May 1, 2004, this report has not been reviewed or endorsed by any oil company or oil industry organization. The information contained in this report represents the author's interpretation and analysis of information that is available to the public. It is not guaranteed as to accuracy or completeness. Although the statements, comments, conclusions, and forecasts prepared for this report are based on logical research, they are presented without any warranty.

Audience- Scholars will shudder at the brevity of my explanations. However, my intent is to reach the lay person - not the academic - so a thousand pardons for my humble erudition. In order to capture the attention of the lay person - the technical stuff has to be interesting. And brief.

Additional Information - The reader is invited to pursue further study by reviewing the information found in the Appendices to this report: Appendix 1 Supplemental Information - presents a sample of corroborative information that may be found on this subject; Appendix

2 Alternative Energy - presents a more detailed discussion of two energy options; and Appendix 3, Definitions - will help to clarify the information presented herein. A partial list of secondary research information resources can be found in the References section.

Use of Terms: Muslim and Islamist - It has been my good fortune to have worked with several Muslims over the years. I have absolutely no desire to disparage those compassionate and devoted Muslims who believe that Islam is the path to Allah. We have much in common. On the other hand, it is most unfortunate that some members of this faith have chosen to believe that Islam provides the moral justification for Jihad as a way of life. In this report, the term "Muslim" simply refers to people who follow this faith. The term "Islamist" is used to refer to those who promote an extremist interpretation of Islam.

Scenarios - The insights presented in this report are based on the analysis of four scenarios. Scenarios are not predictions. Rather, they permit us to make, and then test, a hypothesis. We will then be able to challenge the assumptions, encourage debate about the model, and profile the probable result of our analysis. Scenarios are tools that give our evaluations focus, permit us to deal with the unexpected, and characterize the results of dynamic circumstances.

Chapter 1   CRISIS? WHAT CRISIS?

Truth is reality.

The Secret

Truth is unpopular.

Liberals ignore truth if it challenges their beliefs or mocks their preconceptions. They are more comfortable with an adjusted reality that supports their ideology. Conservatives reject truth if it will force change. They are more secure with intellectual pursuit that confirms established principle.

Neither philosophy will like this report. It undermines doctrinaire wisdom and predicts catastrophic change. Our future reality includes the decimation of our economic condition and the subsequent decline of human culture. All that we have achieved, and all that we aspire to, are in mortal jeopardy.

Oil depletion is a secret. The press typically ignores a problem until it causes a crisis. By then it will be too late. Politicians are reactive animals. They shun the topic of oil depletion because it is political suicide to discuss unpleasant issues. If they acknowledge a pending oil crisis, then we voters will ask embarrassing questions: Why did you let this happen? What do you intend to do about it?

And that will lead to more obfuscation. Neither the press nor our politicians have a clue.

Then there is the issue of corruption. A thorough inquiry into the subject of oil depletion would raise uncomfortable questions. Have our politicians been feeding at the oil money watering trough?

So nobody wants to point the finger. Not Democrats. Not Republicans. Liberals or conservatives. Don't rock the boat. Government reports mislead us with unfounded optimism or ideologically correct misinformation. The officious pressure of conformist procedure ensures government bureaucrats will cover for government politicians and academics will reject meaningful discussion.

Oil, it would appear, is not a politically correct subject.

But eventually the voters will find out about oil depletion and the whole Middle East mess. And then the politicians will have to fabricate some plausible excuse for their failure to act. Blame it on the President. He wants to spend billions on space exploration. We can nail him for his failure to propose a comprehensive energy program. What's more important, we will ask - a make work program for NASA - or heat for your home? Pretty pictures of some planet - or gasoline for your car?

Doesn't our government know how to set priorities?

NO.

But we are suspicious. We who will suffer. Ominous signs are everywhere. Rising gas prices. Talk of shortages. Indefensible terrorist activity. A continuing war in Iraq. A growing suspicion of Saudi Arabia. The threat of conflict in most of the Middle Eastern nations where oil is produced. An oil crisis will dump the world's economy down the toilet. Poverty. Unemployment. High inflation. Kids going hungry. Men desperate for a job. Women struggling with demeaning depravation.

Doesn't anybody care? Why isn't energy a high priority subject on the agenda of every industrialized nation? Why isn't oil depletion an issue for all candidates in the American elections of 2004? Why are European politicians avoiding this subject?

Because if our politicians can keep oil depletion a secret, they don't have to deal with it.

Warning

Fair warning. Oil depletion. It's coming. Sooner or later.

Maybe sooner than later. If Islamic fanatics have their way, the oil spigot will be turned off. So it doesn't matter how much oil sits under the ground in some pool of reserves. What really matters is how much oil can we actually produce? And that takes us back to Saudi Arabia. And Iraq. The world's economy, it seems, teeters on the political stability of these two countries.

Saudi Arabia is supposed to have the largest oil reserves in the world. But it is becoming increasingly clear that this country's reserve numbers are bogus. And by the way, isn't Saudi Arabia struggling with a growing political instability that could devastate its oil production?

Iraq? Coalition troops try to keep the peace. Islamic fundamentalists are determined to kick the infidels out. Democracy will not work in a land torn by cultural chaos. Dictatorship and tyranny are the future. Under these circumstances, how can we possibly believe that Iraqi oil will delay the inevitable decline of world oil production?

Yes. Oil consumers have a problem. Big time.

But hold on. We're getting ahead of ourselves. Before we talk about an Islamist induced oil crisis, we need a better understanding of the basic problem. Oil depletion. Reserves. How much of that black gold there is left for us humans to consume.

Actually we don't know. How much oil there is in the ground. Or how much oil we Homo Sapiens can produce. And those who control the information don't want to tell us the truth. They sell perception. They know reality. They feed us the big picture to make us feel safe.

There's no problem - right? They sustain our current level of knowledge. Or is it ignorance?

I know what you are thinking. Those damn oil companies. Who can believe them?

But guess what? The big five oil companies are forced by SEC disclosure rules to be relatively straight. No. They are not without fault. But the worst liars in this circus of misinformation represent national governments. It doesn't matter if a particular nation is a producer or a consumer of oil. For reasons of selfish best interest, they both fib. Obfuscate. Hide the truth.

Consumer national governments are loath to even discuss the truth because - anyway you look at it - oil depletion is bad news. And national governments do not like to tell their constituents bad news. That's the way political systems work. Democracy or dictatorship. Bad news must be avoided at all costs. It creates anxiety. Conflict. Questions. Opposition. Bad news always generates a pervasive distrust that somehow the existing national government is doing something wrong. And eventually the offending government officials will be blamed for causing the problem. People, after all, expect their national government to protect them from harm or deprivation.

Even if that is impossible.

Producer nations want to pretend that they can go on producing oil - forever. Oil reserve information is a state secret. That permits these national governments to fabricate whatever oil reserve numbers they are pleased to present to a gullible public. Big reserve numbers allow them to borrow big bunches of money from the bankers. It gives them prestige and status. But can we trust any of the reserve information published by - or about - the nations that belong to OPEC (Organization of Petroleum Exporting Countries)? Probably not. For example, the reserve gains these nations claimed in the late 1980s are an obvious fabrication. There has never been an independent verification that the oil they claim to have actually exists. And get this - even though they pump millions of barrels per day, their reserve claims <u>never seem to decrease</u>.

Is there a credibility problem?

Even Canada has succumbed to the temptress of misrepresentation. Canada now claims to have the second largest oil reserves in the world, all based on the inclusion of its oil sands in its reserve estimates. Oil sands? Doesn't this ignore the incredible problems of producing oil from sticky rock?

What can we expect from the IEA? That's the International Energy Agency in Paris. The IEA is an autonomous agency linked with the Organization for Economic Co-operation and Development (OECD). The IEA produces the World Energy Outlook. But wait. Where do they get their data? Unfortunately, it would appear that the IEA uses the bogus data of producer nations in its forecasts. And is their outlook influenced by the political bias of OECD members?

What of the United Nations? Would the publication of a comprehensive report on oil depletion tell us the truth? Or would the study bog down in the sludge of self-serving political posturing?

Our political institutions have failed us. It would appear that being politically correct is more important than exposing the truth.

If we can not get a straight answer from the governments of oil producing or consuming nations, or even quasi-government international organizations, then can we get a reasonable assessment of oil depletion data from the oil companies? The answer is yes. And no. It depends on who you are talking to and what data is being discussed.

First the good news. Oil companies listed on any stock exchange in the United States are required, by SEC and various accounting rules, to make a reasonable assessment of their reserve positions each year. Oil reserves, after all, are an important item on the balance sheet of any oil company. Accuracy counts. As a practice, however, oil companies <u>understate</u> their reserves. They do not want to be in the position of inflating the value of their assets. That would be fraud. So they try to fudge a bit on the low side in order to protect themselves. Most investors are pleased with this arrangement.

Now the bad news. Oil companies do not actually own most of the oil they produce. It is produced under a revenue sharing arrangement with a government agency of the nation under whose land (or water) the oil reservoir is located. The oil company has a license or extraction contract. So when an oil company tells us it has so many barrels of oil in its reserves, that figure may include assets that are derived from existing contracts. If the contract goes away, so does the oil that can be claimed as an asset. In addition, at any given time the oil companies are under contract for only a fraction of the reserves that may be claimed by the nation that actually owns the oil.

So this leads us to only one possible conclusion: there may be less oil on this planet than claimed (which would be devastating); or there may be more oil than anyone realizes (which would be very nice).

We simply do not know.

Or do we?

That's what this report is all about.

Chapter 2 INDUSTRY ANALYSIS

We want to ignore oil depletion. And we will. Until it eats us.

Oil Depletion Forecasts

Interest in the subject of oil depletion has been growing. Here is a representative synopsis of current opinion.

Douglas-Westwood

Douglas-Westwood, Ltd. is a respected oil industry consulting firm based in Canterbury, England. According to Douglas-Westwood's 3rd edition of *"The World Oil Supply Report 2004-2050"*, authored by Dr Michael R. Smith, three fundamentals are strongly evident: increasing oil demand, reducing reserves and a decline in discovery rates[1].

"Global oil demand grew by a dramatic 2.6% in 2003 with the greatest increase from China. This was led by a surge in oil-fired power generation capacity and a 75% jump in sales of passenger vehicles. Although the 2003 demand growth of over 10% in China is unsustainable, it is still expected to remain the highest in the world." ...

"The world is drawing down its oil reserves at an unprecedented rate. Although 99 countries have or can produce significant oil, 52, including the USA, are already well past peak (greater than 5 years) whilst another 16 including the UK, Norway, Australia and China are at peak or will reach it soon. All the remainder will see peaks in the next 25 years."...

"Only Russia, Kazakhstan and Azerbaijan can truly lay claim to significant conventional oil reserves and projects here have been very slow to get off the ground. And although deepwaters are a great opportunity, they are currently responsible for only around 4% of production and are expected to only reach 10% of total capacity."...

"Thus the fundamental conclusions remain that the world's known and estimated yet-to-find reserves and resources cannot satisfy even the present level of production beyond 2020. And just 1% growth in global economic activity increases demand such that a production peak

---

1 *"The World Oil Supply Report"* forecasts likely future oil production for all existing and potential oil producing countries. The report defines the year in which oil supplies will be unable to continue to meet a given global demand. Unlike many of the world's publications on oil reserves, this report is not limited by commercial pressures or political needs. The World Oil Supply Report is used by organizations world-wide. Data is drawn from the 'EnergyFiles' oil and gas supply information system.

occurs as early as 2016. Although the response will be complex, this will ultimately result in a sustained increase in oil prices."...

"A short period of over-supply is forecast, generated by growth from the former Soviet Union and from deepwater (production) ... However, by 2008 all OPEC countries will need to begin to increase production as much as they can to meet even modest demand growth. Iraq is probably the most under-explored country in the world in relation to its productive potential. Once stability is achieved a field development programme must begin with very large infrastructure projects and huge investments. However, whether companies will be willing to take on the political and geological risk, remains to be seen," .

The report stresses the need for rapid action by consuming nations. "All governments must review energy supply security now to produce policies and budgets consistent with impending shortfalls in oil supply in the coming years. Japan and China are competing for Russian oil whilst a number of countries in Western Europe are already facing up to the prospect of energy supply shortfalls and beginning major programmes to encourage renewables. However, Europe, Asia and North America will need a lot more than renewable energy to overcome the long term supply squeeze."

USGS

U.S. Department of the Interior, U.S. Geological Survey
From: USGS World Petroleum Assessment 2000.

"The USGS periodically estimates the amount of oil and gas remaining to be found, and since 1981, the last three of these studies has shown a slight increase in the combined volume of identified reserves and undiscovered resources.

In USGS World Petroleum Assessment 2000, the world was divided into approximately one thousand petroleum provinces, based primarily on geologic factors, and then grouped into eight regions roughly comparable to the eight economic regions defined by the U.S. State Department. Significant petroleum resources are known to exist in 406 of the 1000 geologic provinces.

The U.S. Geological Survey's latest assessment of undiscovered oil and gas resources of the world reports an increase in global energy resources, with a 20 percent increase in undiscovered oil and a slight decrease in undiscovered natural gas. This assessment estimates the volume of oil and gas, exclusive of the U.S., that may be added to the world's reserves in the next 30 years."

"There is still an abundance of oil and gas in the world," said Thomas Ahlbrandt, USGS World Petroleum Assessment project chief. "Since oil became a major energy source about 100 years ago, about 539 billion barrels of oil have been produced outside of the U.S. We now estimate the total amount of <u>future</u> technically recoverable oil, outside the U.S., to be about 2120 billion barrels."

" ... The USGS team believes the largest reserves of undiscovered oil lie in existing fields in the Middle East, the northeast Greenland Shelf, the western Siberian and Caspian areas, and the Niger and Congo delta areas of Africa. Significant new reserves were found in northeast Greenland and offshore Suriname, both of which have no history of production. 'What we did is look into the future and predict how much will be discovered in the next 30 years based on the geology of how it gets trapped,' explains Suzanne D. Weedman, program coordinator of the USGS World Petroleum Assessment 2000. 'We also believe that the [oil] reserve numbers are going to increase.' the USGS report is documented with 32,000 pages of data."

The USGS conventional oil reserve numbers would appear to break down as follows:

World Resources, excluding the U. S. A.	
Undiscovered Oil (as of 2000)	649 Bbl
Reserve Growth (as of 2000)	612 Bbl
Known Reserves (as of 1995)	859 Bbl
Total	2.120 Bbl
Add U. S. A.	143 Bbl
Total World resources, including the U. S. A.	2.263 Bbl

If we correct the USGS Known Reserves data for production since 1995, Total World Resources of conventional oil and NGL as of 12/31/2003 would appear to be 2.047 Bbl (billion barrels).

EIA

Energy Information Administration[2]

From: International Energy Outlook 2003 (IEO2003)

(Italics were inserted by the author for emphasis).

"Worldwide consumption of commercial energy is projected to grow by 58 % over the next two and one-half decades ... (with) much of the growth to occur in the developing world ...

In the IEO2003 reference case projection, world oil consumption increases from 77 mm bpd in 2001 to 119 mm bpd in 2025, *an annualized growth rate of 1.8 %.*

The increases in worldwide oil use projected in the reference case would require an increase of 42 mm bpd over current productive

---

2 United States Energy Information Administration, Department of Energy, www.eia.doe.gov, Provides data on petroleum production, pricing and availability.

capacity. OPEC producers are expected to be the major source of increased production, but non-OPEC supply is expected to remain competitive, with major growth in offshore resources, especially in the Caspian Basin, Latin America, and deepwater West Africa. ..."

From: Annual Energy Outlook 2003

"As has been typical over the past few years, energy prices were extremely volatile during 2002. Spot natural gas prices, about $2 per thousand cubic feet in January, rose to between $3 and $4 per thousand cubic feet by the fall. Average wellhead prices, which are moderated by the inclusion of natural gas bought under contract, also increased over the year. Crude oil prices also rose in 2002, mainly because of reduced production by the Organization of Petroleum Exporting Counties (OPEC) and, to a lesser degree, fears about the potential impact of military action in Iraq. Crude oil prices began 2002 at roughly $16 per barrel and were between $25 and $30 per barrel by the fall. ...

Net imports accounted for 55 percent of total U.S. oil demand in 2001, up from 37 percent in 1980 and 42 percent in 1990. That trend is expected to continue. A growing portion of imports is projected to be refined petroleum products, such as gasoline, diesel fuel, and jet fuel, assuming the future availability of those products in world markets. ...

In nominal dollars, the average world oil price is expected to reach approximately $48 per barrel in 2025.

World oil demand is projected to increase from 76.0 million barrels per day in 2001 to 112.0 million barrels per day in 2020 including projected demand in the former Soviet Union and in developing nations, including China, India, Africa, and South and Central America. *World oil demand, including both conventional and unconventional oil supplies, grows to 123.2 million barrels per day by 2025*. Growth in oil production in both OPEC and non-OPEC nations leads to relatively slow growth in prices through 2025. OPEC conventional oil production is expected to reach 60.1 million barrels per day in 2025, more than double the 28.3 million barrels per day produced in 2001. The forecast assumes that sufficient capital will be available to expand production capacity.

Non-OPEC conventional oil production is expected to increase from 45.5 to 58.8 million barrels per day between 2001 and 2025. A 1.0 million barrel per day decline in production in the industrialized nations (United States, Canada, Mexico, Western Europe, Japan, Australia, and New Zealand) is more than offset by increased production from Russia, the Caspian Basin, Non-OPEC Africa, and South and Central America (in particular, Brazil). Russian oil production is expected to continue to recover from the lows of the 1990s and to reach 10.4 million barrels per day by 2025, 44 percent above 2001 levels. Production from the Caspian Basin is expected to exceed 5.0 million barrels per day by 2025, compared with 1.6 million barrels per day in 2001. By 2025, projected production from South and Central America reaches 6.3 million barrels per day, up from 3.7 million barrels per day in 2001. Non-OPEC African production is projected to grow from 2.7 million barrels per day in 2001 to 6.9 million barrels per day by 2025. ...

USA Total energy consumption is projected to an average annual increase of 1.5 percent. Light-duty vehicle miles traveled are projected to grow by 2.4 percent per year through 2020 and by 2.3 percent per year through 2025. Consistent with recent trends, less improvement is projected for the average fuel efficiency of new light-duty vehicles than in *AEO2002*. New light-duty vehicle efficiency is projected to reach 25.6 miles per gallon by 2020 in *AEO2003* ... and 26.1 miles per gallon by 2025.

USA Total petroleum demand is projected to grow at an average annual rate of 1.7 percent through 2025 (reaching 29.17 million barrels per day), led by growth in the transportation sector, which is expected to account for about 74 percent of petroleum demand in 2025. ... "

IEA

International Energy Agency[3]

From: Oil Supply Prospects

(Italics were inserted by the author for emphasis).

"*Oil reserve estimates are inevitably uncertain* and studies normally report oil reserve estimates as ranges, rather than as point estimates. For example the United States Geological Survey in 1993 reported a range of 2.1 to 2.8 trillion (10^{12}) barrels for worldwide recoverable reserves of conventional oil. Experts differ on these figures; some take a static view, emphasizing geological and statistical issues that lead to a low reserve estimate, and some take a dynamic view, arguing that rapidly advancing technology will help discover more reserves and make a wider range of already known deposits economically recoverable. *Experience in mature oil regions indicates that production builds to a peak when approximately half of the ultimately recoverable reserves has been produced, and then falls away.* The application of new technologies, such as horizontal drilling and 3D seismic analysis, determines the ultimate size of recoverable reserves. It can extend the peak and delay or slow the decline in production. But eventually production falls, given a fixed oil resource. This has been the experience, for example, in the United States.

---

3 International Energy Agency, www.iea.org, Based in Paris, the IEA is an autonomous agency linked with the Organization for Economic Co-operation and Development (OECD). The IEA produces the annual World Energy Outlook.

This approach has been applied on a regional basis. *It indicates that a peaking of conventional oil production could occur between years 2010 and 2020*, depending on assumptions for the level of reserves. Oil production outside OPEC Middle East would peak before OPEC Middle East production *implying a greater reliance on OPEC Middle East supply* between the two peaks. A plateau in oil production for OPEC Middle East of 47.9 Mbl/day has been assumed, rather than a sharp peak, following an IEA study.

... projections for oil production profiles for the world ... (assume) ultimate recoverable reserves of conventional oil of 2.3 trillion barrels. ... Table 1 gives details of supplies for conventional and non-conventional oil. The transition from conventional to non-conventional oil as the marginal supply in 2015 is assumed to raise the oil price from $17-25 ... over the period 2010 to 2015. *The use of non-conventional oil expands rapidly after 2015* as it meets the increase in demand for liquid fuels and compensates for the decline in conventional oil production.

.... To produce large and increasing volumes of oil from non-conventional sources will require many major multi-billion dollar projects. Some *unevenness in supply availability* is possible because of the long lead times required for these big projects and the difficulties in matching supply to demand ... . It is necessary to distinguish fluctuations in the world oil price from its longer term average level. Some *short-term price movements could well arise from supply-demand mismatches*, ... But opinion on the effect of this changeover on ... oil price is mixed."

"A higher view of oil reserves would assume an ultimate stock of recoverable conventional oil of 3 trillion barrels, compared with the lower assumption of 2.3 trillion barrels.... This view postpones the production peak of conventional oil and the associated rise in world oil price to 2020."

Table 1
IEA Oil Supply 1996-2020*

	1996	2000	2010	2020
Million barrels per day (Mbl/day)				
Total Demand For Liquid Fuels	72.0	78.3	94.5	110.1
Total Natural Gas Liquids, Processing Gains and Identified Unconventional Oil	9.3	11.6	15.5	20.6
Conventional Crude Oil				
Middle East OPEC	17.2	20.1	40.9	45.2
World excluding Middle East OPEC	45.5	46.6	38.0	27.0
Total Crude Oil .	62.7	66.7	78.9	72.2
World Liquids Supply excluding Unidentified Unconventional Oil	72.0	78.3	94.5	92.8
Balancing Item - Unidentified Unconventional Oil	0.0	0.0	0.0	17.3

*Assuming a Lower Estimate of Conventional Oil Reserves of 2.3 trillion barrels

To its credit, the IEA does present us with a relatively unambiguous analysis of its data. The stark truth comes out if we are willing to analyze the implications of the IEA's report.

The IEA apparently believes that *conventional crude oil shortages are inevitable.* In order to provide the world with enough energy, non-conventional fuel production will have to increase rapidly after 2014. According to the above Table, by 2020, we humans will have to be producing 17.3 Mbl of un-conventional petroleum liquids per day (6.3 Bbl per year). The IEA has illustrated this shift in the source of oil production in the following chart. This chart also shows two other important points: *non-OPEC oil production has already peaked, and OPEC oil production will peak about 2014 - 2018.*

It's all downhill from there.

Figure 1
IEA: Oil Supply Profiles 1996-2030

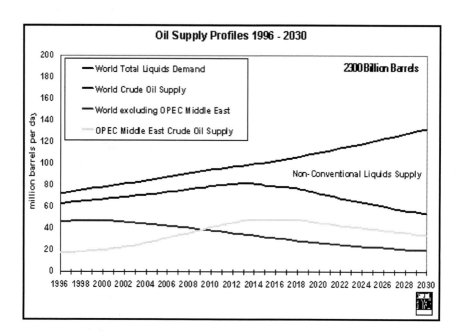

Independent Analysts

Colin Campbell.

October 23. 2002. Professor Colin Campbell, a senior oil geologist and oil executive with forty years of experience in oil exploration and production was interviewed by "From The Wilderness" Editor Mike Ruppert. The complete text can be found at WEB site fromthewilderness.com/free/ww3/102302_campbell.html.

"The key event in the Petroleum Era is not when the oil runs out, but when oil production peaks, especially as demand and population are rising. World per capita oil production peaked in 1979 and has been in decline since. The peak in volume of total world oil production is upon us right now, even as the demand or better said -- the need -- for oil is increasing rapidly.

Several things are a given. First the total remaining conventional oil on the planet is estimated to be around 1 trillion barrels. Second, at present rates (not those of five or 10 years from now), the world is using close to 80 million barrels per day. At the current rate, there would be only enough oil to sustain the planet for another 35 years under the best of scenarios. But the oil that remains is going to be increasingly expensive to produce and it will tend to be of a lesser quality, necessitating higher refining costs, than what has already been used. All of those costs will have to be passed on in the form of price hikes or -- in some cases -- spikes. Oil price spikes invariably lead to recession. The world's economy is based upon the sale of products that are either made from oil or which need hydrocarbon energy (including natural gas) to operate, either via internal combustion or via electricity.

Different regions of the world peak in oil production at different times. The U.S. peaked in the early-1970s. Europe, Russia and the North Sea have also peaked.

> However the OPEC nations of the Middle East peak last.
> Within a few years they -- or whoever controls them --
> will be in effective control of the world oil economy, and, in essence,
> of human civilization as a whole. ...
> Colin Campbell

The majors are merging and downsizing and outsourcing and not investing in new refineries because they know full well that production is set to decline and that the exploration opportunities are getting less and less. ...

Only a fraction of the oil in the reservoir is recoverable because it does not sit in one big cavern down there but in the very small pore spaces between the grains of sand. These grains are coated in water and when it coalesces, it blocks the pore spaces preventing the further movement of oil. ... It is said that recovery has increased from 30 percent to 40 percent thanks to technology and is set to rise ...in the future. But most of this improvement has nothing to do with technology. It is an artifact of reporting. The industry has always made conservative initial estimates ...so reserves naturally grow over time."

Duncan and Youngquist.

Richard C. Duncan and Walter Youngquist, "Encircling the Peak of World Oil Production," Natural Resources Research, (1999):3:219-232, p. 220.

"Oil has formed in the upper approximately 16,000 ft of the Earth's crust since at least as far back as the Cambrian Period, some 550 million years ago .... It is a rich inheritance of highly concentrated solar-derived energy captured by myriad organisms, chiefly algae, and then distilled by geological processes into an energy form that is unequalled by any other energy source in its versatility and convenience in handling. Now, within one human lifetime, one-half of this unique 550 Million Year inheritance will have been spent. The remainder will go very fast." The article predicts that by 2007 the world will peak in oil production at 30 billion barrels per year and by 2020 we will have dropped to 24.6 billion barrels per year.

James Puplava.

From an excellent document prepared by James J. Puplava, Part 1: Hubbert's Peak & The Economics of Oil; available on Financial Sense, March 16, 2002; financialsense.com.

"Hubbert estimated world oil reserves at 1.8 trillion barrels. Since that time and including new discoveries, the estimate has been raised to 2.0 to 2.1 trillion barrels of oil. We now have data on world oil production going back to 1850. Colin J. Campbell of Petroconsultants has made a country-by-country estimate of the world's oil reserves. His estimates match those of Hubbert at 1.8 trillion barrels. With updated data, there has been no significant bulge or dip in the world's production curve as originally estimated by Hubbert. Using production decline curves from known oil reserves, petroleum analysts using Hubbert's methods have now been able to estimate the peak in world oil production. According to these estimates, world oil production will begin to peak between the years of 2004 and 2008. A few of the top geologists, including Colin J. Campbell, think that peak is in 2003. The point to understand is that we are depleting our oil reserves at an annual rate of 6% a year; while demand growth is growing at an annual rate of 2%. In order to simply keep even, the world's oil industry would have to find the equivalent of 8% a year of new oil reserves from new discoveries. This is not happening. The world consumes 76 million barrels of oil a day. This oil is not being replaced.

There aren't any conservation measures like more efficient refrigerators, better gasoline mileage for autos, better insulated homes, or longer lasting light bulbs that could conserve enough energy to make up the difference between demand and supply. For that matter, there are no renewable energy projects on the horizon that could immediately help us to avoid a future energy crisis.
James J. Puplava

There is nothing going on in the Caspian Sea, West Africa, or the South China Sea that would come close to replacing what we are now consuming.

14

Political leaders and the public are totally oblivious to this fact. They are not paying attention. Last year's energy crisis has now been forgotten. The news media explained the crisis in terms of industry price gouging, regulations, taxes, and distribution problems. No one is paying attention to world production declines in the U.S. or elsewhere. This means there is nothing that can be done now in order to avoid a future crisis. It takes years from the time of discovery of new oil to the time it is produced, shipped, and refined and consumed as energy. The oil that is discovered today won't reach the markets for another 8-10 years. An unprecedented crisis is just over the horizon because of inattention and neglect."

Confusion And Reality

It should be obvious that both the USGS and the Department of Energy are under pressure to make sure their respective projections of oil consumption are matched by oil production and reserve availability. Publishing data that shows oil production will be less than oil demand would not be politically correct.

Critics of the USGS World Petroleum Assessment 2000 point out that this survey relies heavily on statistical analysis. The USGS has assumed that if certain types of geological formations in one part of the world have yielded deposits of oil in the past, then it follows that oil deposits will be equally abundant in these formations where they appear in other parts of the world. The USGS has thus identified the formations where oil is likely to be found. These geological structures are well known. Unfortunately, the USGS has determined its reserve estimates by making probability calculations, rather than by drilling holes in the ground. Whether there is, or is not, any oil in these structures is therefore speculative. The USGS also comes under fire for its oil recovery assumptions, which appear to be 30 to 33 percent higher than current oil production experience.

The IEA and the U. S. EIA/DOE appear to rely on statistics published by various trade publications, most notably the "Oil and Gas Journal" and "World Oil". These statistics are, in turn, heavily influenced by the politically motivated reserve claims of the oil producing nations. The reserves routinely quoted by the member nations of OPEC, for example, are highly suspect because even to casual observation it is obvious they have been manipulated.

There is a high level of confusion over the classification of oil reserves. For example, what is the difference between "identified reserves" and "proven reserves"? If proven reserves of 859 Bbl of oil are defined to include all Identified reserves, then total available reserves of conventional oil increase to 1.1 Tbl (Trillion barrels of oil). Total available reserves, including oil that may be found in deposits under the ocean, oil from tar sands and shales, oil that is located under layers of polar ice and oil derived from petroleum liquids, may exceed 3 trillion barrels. That would appear to give us some breathing room in our oil depletion scenario.

But wait a moment. If we accept these most optimistic estimates of oil reserves, is it likely that we will be able to find all of them?

No. Logic infers that a percentage of possible oil reserves will never be found because they exist in formations that are inaccessible - under the ocean, under layers of ice, or in lands plagued by bad weather and hostile cultures.

Can we assume that all found "pools" will be large enough to be of practical value?

No. The cost of extraction will exceed the value of found oil in some percentage of reserves. All reserves are not equally useful. Most of the oil found over the last four years, for example, has come from smaller deposits that will deplete rapidly.

Can we assume that we will be able to extract every drop of oil from all available reserves?

No. Although available technology has dramatically increased the volume of oil that can be recovered, all wells have a finite life span. When abandoned, there will still be oil in the ground.

Can we assume, as some have claimed, that oil from tar sands and shales will replace the declining volume of conventional oil production?

No. Heavy oil must be "thinned" by the addition of condensates (Pentane, for example) in order to manufacture a feedstock that can be refined into petroleum products. Existing supplies of these dilutents are limited. In addition, heavy oil production uses increasingly scarce natural gas to heat the oil so that it will flow from its source rock. Natural gas shortages will curtail production.

How realistic is the EIA assumption that OPEC suppliers will be able to increase conventional oil production from 28.3 million barrels per day in 2001, to 60.1 million barrels per day in 2025?

Given the realities of Middle East politics and projected oil production capacity, this assumption is highly speculative.

How realistic is the IEA assumption that non-conventional liquids production can increase fast enough to make up the difference between increasing demand and declining conventional oil production?

If we humans immediately adopt the recommendations found in this report, it is possible that we can grow our non-conventional liquids production fast enough to compensate for declining conventional oil production. Unfortunately, it is highly unlikely that existing alternative energy programs will be available for mass commercialization before we encounter a depletion crisis.

From the 2003 Update of the ASPO (Association for the Study of Peak Oil) Oil & Gas Depletion Model (Colin Campbell & Anders Sivertsson).

"We are using oil faster than we are finding it and have done so since about 1981. ... Although there remain the eternal uncertainties about the reliability of the data, it appears that the world's oil account has been running a deficit since 1981, as it continues to eat into its inheritance from past discovery."

And consider Colin Campbell's base case for ASPO: "Total world conventional petroleum production from day one to year 2075 (including already produced and yet to be discovered) is estimated at about 1,900 billion barrels, and unconventional oil production (includes natural gas liquids, bitumen, deepwater, arctic, etc.) increases this total to about 2,700 billion barrels. Oil production from all sources is expected to rise to about 83 million barrels per day in year 2010 and then begin a terminal decline."

Going forward, oil companies are going to be under increasing pressure from government agencies to be sure they have accurately stated their estimated reserves. For example:

Thursday, March 18, 2004, AP Biz Wire, seattlepi.nwsource.com
"The Royal Dutch/Shell Group of Cos. ... announced additional cuts to its estimated reserves of oil and natural gas and suggested that more reductions might follow. ... Shell has now reclassified 4.15 billion in reserves that it had carried on its books."   Reserves under contract are carried as an asset on the company's books.

Market Trends

The Peak

Disaster prophets like to point out that once oil production peaks, it will then decline at a rate that is equal to, or greater than, the rate of increase experienced when there was a surplus of reserves. This is the very sharp "peak" of oil production so often seen in various articles on oil depletion. If geology and the mechanics of extraction were the only variables, this would probably be the case. A sharp peak would occur and we would subsequently feel as though the availability of oil were falling off a cliff because of a rapidly growing delta between real demand and available production. However, since neither production nor consumption have historically followed a smooth curve - up or down - and since there is an economic interaction between demand, consumption and production, we should expect the peak of consumption to be characterized by a series of alternating cycles. Periods of shortage will be separated by periods of surplus. Shortages will curtail economic activity. As oil production recovers from a shortage, it will become available to a market where demand has been decreased by the recessionary influence of the previous shortage. Surplus oil decreases prices. Economic theory suggests that the combination of lower prices and surplus oil should stimulate demand - assuming, of course, there are no peripheral events to limit an economic recovery. The recovery will continue until increasing oil demand again exceeds production. At this point, consumption should equal production, plus or minus variables such as available storage, weather conditions, transportation snafus, rank speculation, cultural conflict, labor strikes, political conflict, and so on. If thereafter production should decrease, for any reason, consumption will also be forced down. Decreased oil consumption, along with the associated increases in petroleum prices, will cause a corresponding decrease in economic activity. The cycle will thus be repeated.

We must conclude, therefore, that oil induced recessions, punctuated by periods of increased economic activity, become a distinct possibility as oil production peaks. These cycles could become very severe in magnitude, and the cultural challenges discussed elsewhere in this report will certainly serve to exacerbate their volatility. These cycles may have already begun. We shall certainly see their impact well before 2010.

Real Versus Natural Demand

Boring as economics may be to some readers, it is very important for us to understand that at the "peak" of oil production, and beyond, there will be a growing divide between demand, consumption and production. Indeed, demand will split into to components: Natural Demand and Real Demand.

Natural demand quantifies the theoretical demand for oil that would exist if there were no impediments to consumption. China, for example, currently has a projected average annual natural demand of 3.6 percent per year. That is how fast China's natural demand for oil would grow each year (on average) if there were no restrictions on consumption or economic growth for the period 2003 through 2022. Natural demand is not merely a theoretical number. It reflects the economic and cultural expectations of a consuming nation. If oil production fails to keep up with natural demand, then there will be social discontent because consumers have to reduce their expectations. Oil shortages will restrict consumption as we approach the peak of oil production. Consumer confidence will subsequently decline, and this decline may trigger a recessionary trend in the national economy.

Real demand, by contrast, measures how the demand for oil ebbs and flows according to changes in price and the economic health of consumer nations. Growing economies and lower prices stimulate additional demand. Recessionary economies and higher prices reduce demand. As we approach the peak of oil production, cycles of shortage and surplus will increase the volatility of real demand. During shortages, consumers will be compelled to reduce their demand to match the actual availability of oil - as happened in 1973. Real demand will be forced down until consumption equals production. But what happens when there is a subsequent surplus? The growth of real demand will be constrained by the health (or lack thereof) of the economy. In up cycles, we can expect real demand to trail available oil production until the economy recovers. Thus we can expect economic down cycles caused by oil shortages and higher prices to happen very fast. The psychological impact will be traumatic. Up cycles based on a surplus of available oil and lower prices will take more time to develop. Consumer confidence will take time to recover.

And so. Why is this discussion important? Because the cycles of consumption and production discussed above will increase the delta between natural demand and real demand. With each passing year, this delta will increase. The larger the delta, the greater the social discontent. Combine the volatile economic hardship of oil shortages with the growing delta between natural and real demand, and it becomes obvious that after oil production peaks, there will be a corresponding increase in social discontent.[4]

The Price of Oil

Pumping Oil

We humans have been lucky. Although the price of oil has been volatile, there has been little upward momentum since 1979. From here, however, oil prices will trend higher. Sure. There will be intermittent periods of declining prices. But don't be fooled by temporary illusion. The average year over year price of oil will increase at an accelerating rate throughout our forecast period (2003 - 2022).

Pumping oil from the ground is not like pumping water from a barrel. To recover the oil, it must be encouraged to drain through layers of sand and rock toward the foot of the well. Oil collects in rock formations (the reservoir) where the individual droplets of oil may be trapped in the very small pore spaces between the grains of sand.

Only a fraction of the oil in the reservoir is recoverable. If the grains of sand or shale are coated with water, it frequently coalesces, blocking the pore spaces and preventing the further movement of oil. The lighter the oil (higher API) the more readily it will flow toward the foot of the well and recoveries of 50 to 60 percent are possible over time[5]. Heavier oils will take longer to make this migration and recoveries may be in the 30 to 40 percent range. Enhanced oil recovery encourages this migration through the use of steam, water or gas which are pumped into the reservoir to free up the pore spaces and/or thin out the oil. The resulting increase in production is sometimes referred to as "Reserve Growth", because the original estimate of reserve capacity has been increased through the use of enhanced recovery technology. The productivity of an oil field can be further enhanced by drilling more wells, including horizontal bores that actually traverse the oil bearing strata.

When a new well is proposed, petroleum engineers will estimate how much it will cost to drill the well and how much it will cost to connect that well into an existing oil collection system. Initial production will usually be substantially higher than later production because the well pump is able to capture nearby deposits. Over time, production will decline because of pressure loss, water encroachment,

---

4 It should be pointed out that as the years go by, natural demand will decline as people adjust to a more austere lifestyle. Never-the-less, there will always be a growing delta between unmet natural demand and a real demand that has been curtailed by oil shortages, higher prices and subsequent economic decline.
5 That means 30 to 40 percent of the available oil reserves in a given field can eventually be recovered. For heavier oils, this rate of recovery must inevitably decline to 50 or 60 percent.

the increasing distance the oil must travel to reach the foot of the well and the depletion of the reservoir. If daily operating expenses are reasonably constant, the cost per barrel pumped to the surface must increase as the volume of recovered oil declines. When the cost of production equals the market value of a barrel of oil, production stops.

The optimists claim that we do not have a depletion problem. New technology such as 3D seismic imaging, directional drilling, "smart" drill bits, improved sensors and so on, will increase the amount of oil we can get from any given well. If new technology will decrease the cost of finding and extracting oil, then there will be a resulting net increase in our reserves. When we move to a new energy system, there will still be oil in the ground.

Glenn R. Morton

"There will always be oil in the ground. There is many times more oil in the ground than we have burned. The problem is that it is too dispersed to be extracted economically. No one is going to spend $10 to get $9 of oil. Occasionally we do that in the oil industry, but I assure you that heads roll at those times. What limits the amount of oil we find is economics and the laws of physics. ...

The oil left in the ground unused can not be moved. Oil is often found in sandstone. It is actually found in the pore spaces between the sand grains. Thus the sand is like a sponge. We can measure the oil saturation and the water saturation of a sandstone reservoir. However, when we pump oil out of the ground, there is an irreducible saturation beyond which the oil won't move. ....

As to finding more oil .... In the 1960s, people working the North Sea found fields of 1-2 billion barrels each. Today the average field size is around 30 million barrels--almost a 100 fold reduction. (In the Frio trend of Texas 80 years ago, they found billion barrel fields--in the 1990s the average field size found was 4 million barrels). Now, if that 30 million barrel field is far from infrastructure (pipelines), it takes almost as big a platform to produce the 30 million (bl field) as it does to produce a billion barrel field. Lets say that a platform costs $500 million, which isn't far from the truth in some fields. For a billion barrel field that is fifty cents per barrel cost. For a 30 million barrel field it is $16 dollars per barrel. and that is just for the cost of the platform. That doesn't count the cost of running the platform, paying the salaries ..., paying the refinery costs, the pipeline costs, the wellhead processing costs, and don't forget royalties ... etc. on top of that. ... "[6]

---

6 Glenn Morton, is a Geophysicist in charge of reservoir simulation for a major oil company. Used with permission.

Matt Simmons

According to Matthew Simmons, an oil industry consultant and investment banker, worldwide Public Exploration and Production companies spent $410 Billion Dollars between 1996 and 1999 to merely maintain a flat production rate of just under 30 Million barrels of oil per day. The Big Five: Exxon, Shell, BP, ChevronTexaco, and Total, spent 150 Billion dollars between 1999 and 2002 to make an almost insignificant gain in production from 16 to 16.6 million barrels per day. Exxon, Shell, BP and ChevronTexaco spent over $40 Billion dollars between the first quarter of 2002 and the first quarter of 2003 on exploration and production. Despite this expenditure, production actually dropped from 14.611 to 14.544 million barrels of oil equivalent per day. "One of the other interesting mantras of the last decade", he said, "was that technology had eliminated dry holes. Well, we never came close to making the dry hole obsolete. The reason dry holes dropped so much is we drill far less wells. We also stopped doing most genuine exploration."

Price Pressure

So there is no magic spigot that can be turned on whenever we need more oil. Production from existing wells is limited by the laws of physics and chemistry as well as plain old economics. And spending big bucks for exploration may not produce a significant increase in available oil reserves.

In the 1970's, the price for a barrel of oil went from an average of $1.80 in 1970 to $30.03 in 1979. According to economic theory, if production had been elastic, higher prices should have attracted more production, thereby driving down the price. But that did not happen. There was a relatively insignificant increase in production. In fact, the increase in production from 17.8 Bbl in 1970 to 24.5 Bbl in 1979, was just enough to meet increasing consumer demand. Over a ten year period, a 1,568 percent increase in price only stimulated a 37.6 percent increase in production! Obviously oil production, excluding unused capacity that may be available in periods of weak demand, is *inelastic*. The primary producer nations are now able to set the price of a barrel of oil to whatever value they deem prudent. Thus far, that price has been set to maximize the dollar revenue that can be derived from selling an optimum number of barrels of oil. Nations like Saudi Arabia know that if they raise the price too high, consumer nations will fall into an economic decline, thereby reducing the number of barrels of oil that can be sold. So they have usually adjusted the price to provide a maximum return based on a healthy consumer economy.

It is becoming more expensive to find and produce oil. We have not found a really big - and easy to exploit - pool of oil for a number of years. With the exception of the Middle East, and some fields in the southern latitudes, newer discoveries - on land and in shallow water -

tend to be on the smaller side. They will be depleted rather quickly and the ratio of production cost to recovered oil is becoming less and less advantageous. As we are forced to look for oil in deep water and polar regions, this cost ratio will become even more expensive. Oil drilling costs versus projected oil recovery make it impractical to drill for oil in smaller deep sea and polar fields. We can only justify recovery operations in the larger finds.

The quality of found oil will decline. Yes. There will be exceptions. But the trend will be toward the production of heavier oils, as well as oils contaminated with chemicals, heavy metals and salt water. Taken together, these factors will increase the cost of refining oil. Not only will the refining process be more difficult, it will also require the addition of more solvents and other chemicals to complete the chemical reaction that "cracks" oil into useful products.

Political volatility will increase the cost of finding, producing and transporting oil. Nigeria, Venezuela and Iraq are current examples of this reality. Future political instability in Iraq and Saudi Arabia will have a critical impact on the price of oil. These challenges will increase the cost of oil leases, disrupt operations and hamper transportation.

It should be clear - the days of relying on "Texas Sweet Crude" or "Saudi sweet" for our oil needs are waning. Exploration, production, and refining operations will become increasingly more expensive.

The price of oil will go up.

Reserves Will Decline

From the BP Statistical Review of World Energy, June 2002: "Proven oil reserves are growing as new discoveries and increases in recovery rates due to technical advances are outpacing production, but as an annual percentage reserves are not rising as fast as production." In other words, we are using oil faster than we are finding it. In all my months of research, I found nothing of substance which could be used to refute this fact. Aside from short term speculative euphoria that may erupt with the discovery of a new field, there will be a continuing decline in overall average reserves.

Most oil industry insiders will acknowledge that oil is becoming increasingly difficult to find and more costly to exploit.

On April 12, 2002, Fadel Gheit, an analyst at Oppenheimer & Co. commented that oil companies are running out of good drilling prospects[7]. Oil companies are caught between increasing exploration and development costs, increasing production costs and declining exploration results. Rising gas and oil prices have actually masked a

---

7 Oil mergers' focus: grow production, CBS MarketWatch, April 12, 2004

general weakness in the industry. J. P. Morgan reported that a lack of growth opportunities would lead to more industry consolidation as companies attempt to shore up their petroleum reserves.

So we have a serious problem. If reserves are not rising as fast as production - a fact backed by industry consensus - it is only a matter of time before production must also inevitably decline.

The issue is no longer - "will production decline". The key issue and only point of disagreement is "when"?

Derivative Factors

We must recognize that in addition to production and exploration issues, oil exporting countries are now able to control the price for providing us with an increasingly scarce commodity. In other words, the world oil market has transitioned from a market controlled by consumer demand to one controlled by producer capacity. Producer nation motivation and cultural stability has become a factor in determining the availability and price of oil.

> In the final analysis, therefore, corporate behavior, government action, cultural stability, economics, legal agreements, geography, weather, crude oil transportation, military diplomacy and the always potent combination of religion and politics are as important as geology in developing oil production forecasts

Call these the *derivative* factors of doing business on a global scale. Each poses a potential disruption to the flow of oil. As a result, proven or identified oil reserves are less important than accessible oil reserves - the oil that can actually be produced without disruption. If we want to understand how much oil will be available over the next 20 years, we must consider how these derivative factors are likely to impact oil production and transportation.

Big Oil is painfully aware of the derivative challenge.
For example -

ChevronTexaco

ChevronTexaco.com

Source: 10Q 5/9/2003 Q1

"Upstream. Changes in exploration and production earnings align most closely with industry price levels for crude oil and natural gas. Crude oil and natural gas prices are subject to external factors, over which the company has no control, including product demand connected with global economic conditions, industry inventory levels, weather-related damages and disruptions, competing fuel prices, and the regional supply interruptions that may be caused by military conflicts or political uncertainty. Community unrest has disrupted the company's production in the past, most recently in Nigeria and Venezuela. The company continues to monitor developments closely in the countries in which it operates. Longer-term trends in earnings for this segment are also a function of a range of factors in addition to price trends, including the company's ability to find or acquire crude oil and natural gas reserves and efficiently produce them.

During 2002, industry price levels for crude oil trended upward from the $20 per-barrel level at the beginning of the year to about $30 in December and continued rising in the first quarter 2003. Average worldwide industry prices for crude oil and natural gas in the first quarter of 2003 were significantly higher than the same 2002 period and were the major factors in this segment's higher earnings between periods. In the first quarter 2003, the average spot price for West Texas Intermediate (WTI), a benchmark crude oil, was about $34 per barrel, compared with about $21 in the year-ago period, and peaked at about $37 per barrel in mid-March. ... The higher prices for crude oil in early 2003 in part reflected the geopolitical uncertainty in Iraq and Venezuela. ...

The company's equity production was marginally lower in the first quarter of this year as a result of civil disruption late in the period in Nigeria and during most of the quarter in Venezuela. ....

Production was restored in Venezuela before the end of March to levels that were in place prior to the nationwide labor strike. After improved security was in effect in Nigeria, much of the production that had been shut in was restored by mid-April.

The expected production level in 2003 and beyond is uncertain, in part because of production quotas by the Organization of Petroleum Exporting Countries (OPEC) and the potential for local civil unrest and changing geopolitics that could cause production disruptions."

ExxonMobil

exonmobil.com

Source: SEC 10K for 2002

3/26/2003

"The operations and earnings of the corporation and its affiliates throughout the world have been, and may in the future be, affected from time to time in varying degree by political developments and laws and regulations, such as forced divestiture of assets; restrictions on production, imports and exports; price controls; tax increases and retroactive tax claims; expropriation of property; cancellation of contract rights and environmental regulations. Both the likelihood of such occurrences and their overall effect upon the corporation vary greatly from country to country and are not predictable."

Total, S.A.

total.com

Source: Amendment No. 2 to Form F-3

8/6/2003

"A significant portion of our oil and gas production occurs in unstable regions around the world, most significantly Africa, but also the Middle East, South America and the Far East. Approximately 28%, 18%, 9% and 7% of our 2002 production came from these four regions respectively. In recent years, a number of the countries in these regions have experienced varying degrees of one or more of the following: economic instability, civil war, political volatility, violent conflict and social unrest. In sub-Saharan Africa, each of the countries in which we have production has recently suffered at least four out of five of these conditions. The Middle East in general has recently suffered increased political volatility in connection with violent conflict and social unrest. A number of countries in South America where we have production and other facilities, including Argentina and Venezuela, have suffered from economic instability and social unrest and related problems. In the Far East, Indonesia has suffered the majority of these conditions. Any of these conditions alone or in combination could disrupt our operations in any of these regions, causing substantial declines in production. Furthermore, in addition to current production, we are also exploring for and developing new reserves in other regions of the world that are historically characterized by political, social and economic instability, such as the Caspian Sea region where we have a number of large projects currently underway. The occurrence and magnitude of incidents related to economic, social and political instability are unpredictable. It is possible that they could have a material adverse impact on our production and operations in the future. ......

We have significant exploration and production, and in some cases refining, marketing or chemicals operations, in developing countries whose governmental and regulatory framework is subject to unexpected change and where the enforcement of contractual rights is uncertain. In addition, our exploration and production activity in such countries is often done in conjunction with state-owned entities, ... where the state has a significant degree of control. Potential intervention by governments ... can take a wide variety of forms, including:

- the award or denial of exploration and production interests;
- the imposition of specific drilling obligations;
- price and/or production quota controls;
- nationalization or expropriation of our assets;
- cancellation of our license or contract rights;
- increases in taxes and royalties;
- the establishment of production and export limits;
- the renegotiation of contracts;
- payment delays; and
- currency exchange restrictions.

Imposition of any of these factors by a host government in a developing country where we have substantial operations, including exploration, could cause us to incur material costs or cause our production to decrease, potentially having a material adverse effect on our results of operations, including profits."

Royal Dutch Petroleum Company
Royal Dutch/Shell Group
Source: SEC Form 20-F for the Fiscal Year Ended December 31, 2001
"(Royal Dutch/Shell) Group and associated companies are involved in the exploration for, and production of, crude oil and natural gas operate under a broad range of legislation and regulations that change over time. These laws and rules cover virtually all aspects of exploration and production activities, including matters such as land tenure, entitlement to produced hydrocarbons, production rates, royalties, pricing, environmental protection, social impact, exports, taxes and foreign exchange. The conditions of the leases, licenses and contracts under which oil and gas interests are held vary from country to country. In almost all cases the legal agreements generally have in common that, they are granted by or entered into with a government, government entity or state oil company, and that the exploration risk practically always rests with the oil company. ... "

In the section titled "RISK FACTORS

"The Group and its businesses are subject to various risks relating to changing competitive, economic, political, legal, social, industry, business and financial conditions. These conditions are described below and discussed in greater detail ... (in the Annual Report).

Price fluctuations
Oil, natural gas and chemical prices can vary as a result of changes in supply and demand for products, which may be global or limited to specific regions and influenced by factors such as economic conditions, weather conditions or action taken by major oil exporting countries.

Currency fluctuations
The Group is present in more than 140 countries and territories throughout the world and is subject to risks from changes in currency values and exchange controls.

Drilling and production results
The Group's future oil and gas production is significantly dependent on successful drilling and well development. There are risks in this process in interpretation of geological and engineering data, project delay, cost overruns and technical, fiscal and other conditions.

Reserve estimates
... Oil and gas reserves cannot be measured exactly since estimation of reserves involves subjective judgement and arbitrary determinations and is dependent, amongst other things, on the reliability of technical and economic data.

Loss of market
Group companies are subject to differing economic and financial market conditions in countries and regions throughout the world. There are risks to such markets from political or economic instability, as well as from industry competition.

Environmental risks
Group companies are subject to a number of different environmental laws, regulations and reporting requirements. Costs are incurred for prevention, control, abatement or elimination of releases into the air and water, as well as in the disposal and handling of wastes at operating facilities. Expenditures of a capital nature include both remedial measures on existing plants and integral features of new plants.

Physical risks
The Group's assets are subject to risk from operational hazards, natural disasters and expropriation of property.

Legislative, fiscal and regulatory developments
The Group's operations are subject to risk of change in legislation, taxation and regulation. For exploration and production activities, these matters include land tenure, entitlement to produced hydrocarbons, production rates, royalties, pricing, environmental protection, social impact, exports, taxes and foreign exchange."

28

BP

bp.com

Statistical Review of World Energy, June 2002

"Concerns about reserve sufficiency are more related to political worries as OPEC and FSU (Former Soviet Union) countries between them control over 80% of proven reserves of (conventional) oil and gas. At end 2001 OPEC had 78% of oil and 45% of gas reserves and FSU had 6% of oil and 36% of gas."

Source: BP Website; 8/19/2003

"Conflict is part of the history of Azerbaijan, Georgia, and Turkey. Each country has suffered periodically from war, unrest and ethnic tension, including in the recent past.

BP's ethical conduct policy states: 'Before we make major investments in a new area, we will evaluate the likely impact of our presence and activities. These assessments will consider the likely impact of major developments on local communities and indigenous peoples, local infrastructure and the potential for conflict and its implications for security.'

All ... companies participating in the projects have considered the risks of future conflict and the means to minimize the potential for conflict. The projects' sponsors are aware of the emerging literature on business and conflict, particularly on the relationship between extractive industries and conflict. The Regional Review, together with the Environmental and Social Impact Assessments and internal country and project risk assessments, have helped to evaluate the likely impact of their ... activities. ...

... In a region where conflict has been prolonged, a number of intergovernmental, bilateral, and non-governmental organizations have developed programs to promote economic and social development. EU, World Bank and NGO initiatives are seeking to promote economic prosperity and integration, build confidence between communities, encourage learning from other conflict resolution processes, and reduce economic and social vulnerability.

While prime responsibility for preventing and resolving violent conflict rests with governments, it is increasingly recognized that business can support positive action by working in partnership, promoting economic growth in which benefits are widely distributed, and influencing conflict through the way in which operations are conducted and through social and community investment. ...

In action too, the project sponsors are working to reduce conflict and regional tensions. For example, the projects aspire to contribute to the reduction of tension and conflict potential through positive local engagement. These contributions should be in the form of care for company employees and their families, new investment in community projects, the avoidance of harm to the livelihoods of people living near project facilities and the pipeline route, and transparent, even-handed treatment of land acquisition and other community issues. More generally, the way in which the projects create wealth and employment, and the ways in which they contribute to the protection of human rights, the development of civil society and efforts to combat corruption, are all relevant to conflict prevention and resolution. ..."

Bogus Projections

We frequently see statements that existing oil reserves will last for so many years *at current rates of consumption*. That assumption - "current rates of consumption" is totally bogus. The demand for oil surges and recedes like the waves of the ocean. It is never motionless. Although the market for oil is a growing market (demand is increasing) the pace at which this demand is growing varies according to the economic health of the consuming nations. Further more, if production falters, we are not only concerned with the fact that consumption can not exceed production, we must also consider the delta difference between increasing natural demand and declining production. It is assumed that if there is a surplus of oil production capacity, then real demand could increase as fast as the economies of the consumer nations. Unfortunately, as we near the peak of oil production, oil shortages will force a decrease in consumption and the price of oil will go up. Higher prices and recessionary forces will drive real demand down until oil production recovers.

What does this all mean? It simply means that if you hear someone claim that we have X number of years of oil left at current (rates of) consumption, the statement would be true if: natural demand (the amount of oil that would be consumed if it were in surplus and available at an affordable price), real demand (natural demand less the impact of higher prices and economic health), consumption (the actual amount of oil that is used), and production (the actual amount of oil that is produced), never change for X number of years. Since there is absolutely no way that will happen, the statement is inherently bogus.

30

Chapter 3 DEPLETION DILEMMA

Many readers will ignore careful analysis in preference to preconceived notions and imbedded ideology. Too bad. Reality ignores intellectual vanity.

What is the Truth?

If we were to list the most important issues facing humanity, oil depletion has to be in the top three. The economic and cultural destiny of mankind is inexorably tied to the availability of oil. It is the world's primary source of energy and virtually the only source of mobile (vehicle) energy. Food production relies heavily on petroleum for fertilizer, equipment fuel, transportation and processing. Petroleum provides the feedstock for thousands of manufactured products. Oil accounts for the largest value of commodity transactions on the world's exchanges. It is impossible to address the problems of famine without abundant oil for cultivation, transportation, processing and distribution. Oil shortages will increase unemployment, poverty, cultural disintegration and political chaos. The depletion of the world's oil reserves will lead, indeed have already led, to international political confrontation over territorial rights and producer nation alliances. Oil is being exchanged for weapons. Consumer nations are being encouraged to ignore the terrorist activity of certain producer nations in exchange for access to oil reserves. And finally, this planet can not support the health and welfare of 6.3 billion people with declining supplies of oil.

Unfortunately, despite the dire need to address the twin issues of oil depletion and production, there is almost a criminal lack of reliable information on these subjects. Published reserve estimates have a questionable value. They must be carefully examined for error, intentional misstatement and imprudent logic. There are four fundamental reasons:

1. There is no compelling reason to tell the truth. It would appear that reserve estimates made by the governments of individual oil producer nations are intentionally inflated to make them look prosperous to potential investors and debt holders. Since oil in the ground makes good collateral for financial agreements, the more oil they can claim to have, the better their credit. In addition, certain OPEC members apparently inflated their reserves by over 280 Billion barrels from 1985 to 1989 because their individual oil production quotas were pegged to their declared reserves. The larger the reserves they could claim, the more oil they could produce under the OPEC oil production quota system. There is no record of any discovery or assessment that would justify this

incredible increase in the oil reserves of these nations. In addition, the stated reserves of these nations never seem to decrease, even though they produce millions (or billions) of barrels of oil each year.

On the other hand, oil companies - for their part - appear to understate their reserves. This gives them a buffer of oil assets so that if they have a bad year of exploration and development they can still show a reserve increase by simply revising their estimates upward. This anomaly of the oil industry is being addressed by new accounting regulations that took effect at the end of 2003. Oil reserve asset claims made by companies listed on American stock exchanges will have to be more closely matched to regulatory agency definitions of proven reserves. Oil company reserve numbers are therefore driven by these reporting requirements, rather than by how much oil is actually in the ground.

However, individual countries - like Saudi Arabia or Angola - are not bound by these new reporting regulations. For them, it will be business as usual.

What is the Truth?
Nobody really knows.
As a practical matter - precision is impossible.

---

2. No one really knows. It's true. There is no precise way to measure the amount of oil that lies under the ground. Even with advanced seismic and computer technology, there is a margin of error because of all the variables that will impact future production. How viscous is the oil? What other liquids will enhance or impede production? How deep and wide is the deposit? Where is the best place to place the well bore (or bores)? What are the characteristics of the oil bearing strata? How much additional production can be achieved through enhancement techniques? Could the legal nature of the drilling lease force a curtailment of production? Will local cultural conflict decrease production? Can we assume the continuity of the oil transportation infrastructure? The list of variables goes on and on. So we are left with estimates. To be sure, they have been made by people who have an excellent knowledge of oil geology and the mechanics of upstream operations. But oil reserve estimates are always based on a set of assumptions. And assumptions are - well - assumptions. For example, a few years ago there was a lot of excitement about oil finds in the Caspian basin. Estimates ranged upwards to over 200 Billion barrels. Yet further exploration and development has revealed that these deposits are not in big pools. Instead, they are located in marginal crescents that

will be much more difficult to exploit. So how much oil is in the Caspian basin? A lot. But we will not have a firm understanding of the potential reserves until we drill holes in the ground.

3. Quality. Oil in the ground is not just oil. It may be accompanied by water, sulfur, salts and other contaminates. The reservoir may contain natural gas or natural gas liquids. It may have a high viscosity[8] (heavy oil), sulfur content, or contain insufficient hydrogen, making refinery operations more difficult. Taken together, a billion barrels of reserves may, or may not, be a <u>useful</u> billion barrels of reserves.

4. Definitions. Although organizations such as the Society of Petroleum Engineers and the American Petroleum Institute have been joined by other organizations and companies to create a set of standardized definitions for the Oil and Gas industry, their use is voluntary. Within the halls of national governments, bureaucrats frequently espouse their own politically motivated oil reserve definitions. The broad definition of reported reserves used by many countries is based on the estimated volume of petroleum that can be recovered using conventional extraction technology. Estimated reserves in proven fields that have been carefully documented are combined with informal estimates for those fields which have been identified (but not fully documented), and areas where it is believed that it may be possible to find oil. These estimates frequently assume that all reserves - proven, identified or possible - are equally viable, thus ignoring extraction costs and field characteristics which may render them commercially useless.

Thus we start our quest for truth on a very dubious foundation of bad data. We shall have to meditate often and long.

The discussion thus far should be sufficient to make us wary of any reserve estimates. There are people within the oil industry who will say that existing estimates are too low. At the same time, there are those who will point out that some of the claimed reserves do not actually exist. Since much of the relevant data is treated as proprietary information by national governments, it is difficult to sort out fact from fiction.

Never-the-less, we have enough hard evidence and antidotal information to make a realistic assessment of oil depletion. Given the assumptions used, there is an 80 percent chance that one of the three

---

8 The viscosity of any fluid is defined as its resistance to shear or flow, and is a measure of its adhesive/cohesive or frictional properties. The internal molecular friction of oil causes the molecules to adhere to each other. The higher the viscosity, the greater the drag of this friction.

crisis scenarios depicted in this report (or a variation thereof) will be a reasonably accurate depiction of future events.

Each scenario should be treated as a hypothesis. As we move forward and obtain better information, we can prove, disprove, or improve the elements of the hypothesis. In addition, the impact of each crisis scenario on individual national economies needs further work by some very thoughtful economists who know how to construct the relevant impact formulas using tested associative data.

This stuff has graduate thesis written all over it.

Hopefully, there is someone with a very complex econometric model of the world economy on a very fast computer that can be used to improve on my humble forecasting efforts.

Playing The Numbers Game

The people who calculate oil reserves seem determined to confuse us. Some use overly optimistic estimates. Others calculate reserves using very conservative criteria. And some just plain guess[9].

One of the more technically complex methods is to calculate how much oil is recoverable using F5, Mean and F95 odds. The United States Geological Survey (USGS), for example, quotes recoverable oil from U.S. onshore Federal Lands as follows: there is a 95 percent chance (F95) that we could recover 4.4 Bbl of oil, there is a 50/50 chance (Mean) that we could recover 7.5 Bbl of oil, and there is a 5 percent chance (F5) that we could recover 12.8 Bbl of oil from these deposits. However, if oil is $18 per Barrel, then it would be economical to extract only 1.6 Bbl of oil. If the price of oil rises to $30 per barrel, then it would be economical to extract 3.3 Bbl of oil. We can presume that if oil increases in price to the $40 - $50 range, then perhaps it would be economical to extract the 4.4 Bbl of the oil that the USGS is 95 percent sure actually exists.

Frankly, F95, Mean and F5 simulations are just a way of playing with numbers. My experience with similar statistical exercises is that they generally make little contribution to our pursuit of the truth. The USGS generally uses the Mean estimated recoverable oil reserves when it publishes data for our consumption. In other words, given the parameters discussed in the above example, we may be able to extract up to 7.5 Bbl of oil from onshore Federal Lands at some unknown cost per barrel. Russian and Saudi Arabian oil ministers, by contrast, have been accused of using the F5 estimates - which means there is a 5 percent chance that all that oil they claim to have actually exists.

---

9 Additional oil reserve, production, consumption and depletion information can be found by visiting the Web sites found in the References section of this book.

There is a better way to make these estimates. Oil recovery costs money. The more difficult it is to recover, the higher the cost. As oil wells deplete, the cost per each additional recovered barrel of oil must eventually go up. Continued production from a partially depleted well requires the use of recovery enhancement techniques (such as pumping sea water into the oil reservoir in order to get the oil to flow to a location where it can be pumped out of the well). These enhanced production techniques cost money.

Thus we must tie escalating extraction costs to the available information we have on each pool of oil. We need to know how much oil is *economically* recoverable at specific cost levels per barrel. Oil companies must compete against other oil suppliers. The price per barrel of oil on the world market determines the economic viability of every well they have in production. Extraction costs, plus well site processing, storage and transportation to an international port, must be less than the world price for a barrel of oil if they are to make a profit. If the world price is $25.00 per barrel, then these costs would have to be less than $25.00. If processing, storage and transportation costs for a given oil field are $10.00 per barrel, then the extraction cost had better be less than $15.00 per barrel. However, if world oil prices go up to - say - $35.00 per barrel, then it becomes feasible to spend up to $25.00 per barrel for production enhancement and $10.00 for processing, storage and transportation in order to get the most out of each well in this field.

Obviously, as the price of oil increases on the world market, it becomes economically feasible to spend more money on production enhancement. That holds not only for existing wells, but also for drilling additional wells within a given field. Hence, at each higher price point, the economically recoverable oil reserves of any reservoir increase.

In making the oil reserve estimates for this report, it has been assumed that additional oil becomes technically and economically recoverable as the world price increases. It works this way. If the world price per barrel were to stay below $25.00 per barrel, then we have about 700 Bbl of oil that could be economically recovered from known and to be found reserves. At $35.00 per barrel, economically recoverable world oil reserves increase to an estimated 950 Bbl. Given the significantly higher oil prices that were generated by the oil depletion model used in the following Chapters, we humans can expect to eventually recover up to 1.4 trillion barrels of conventional oil and up to 740 billion barrels of unconventional oil.

We must be aware, however, that for any individual well or field, the cost of recovery for each additional barrel of oil accelerates as the field is depleted. Enhanced recovery techniques add cost to operations while the production volume is decreasing. The cost for each barrel of oil can only go one way. UP.

There is, however, an upper limit to the world price for oil. At some point, we humans simply can not afford it. We have other things we need to buy - food, clothing, shelter, etc. Oil based product purchases, like gasoline, must compete for our limited income. So the "we can increase our oil reserves if we are willing to pay higher extraction costs" argument begins to deteriorate as the price of oil increases.

There is an upper limit to how much we can afford to pay for oil. Beyond this limit, the rate of depletion will actually decrease as consumers are forced to allocate their limited income among other necessities.

In addition, there are limits to how much additional oil we can coax out of the ground and the speed at which it will travel to the foot of the well. If oil field engineers are not careful, the liquids or gas being used to enhance recovery may actually find channels that by-pass the oil they are trying to drive toward the foot of the well. Geology, physics and chemistry thus limit the additional volume of oil that can be gained by spending additional money on production. The older the well or field, the more difficult it will be to increase or even sustain the volume of oil being recovered.

There is an upper limit on the amount of oil that can be recovered at any cost. Furthermore, the use of recovery enhancement techniques may only yield a marginal increase in the rate at which the oil will flow to the wellhead. For many of the world's oil fields, we have already reached that limit, so throwing more money into recovery enhancement may only yield marginal increases in the quantity of available oil.

Remember 1973 to 1982? From 1973 to 1974, acute oil shortages and OPEC action increased the price for a barrel of oil by 252 percent ($3.29 to $11.58). Oil production increased by less than 1 percent (21.679 Bbl to 21.735 Bbl). OK. So it takes awhile to increase production. By 1982 (10 years later) the price of oil had risen to an annual average of $31.76 per barrel. But production, however, actually decreased to 21.252 Bbl in 1982.

An 865 percent increase in price had not stimulated any increase in production!

It should be obvious that oil prices and oil production respond to different metrics, and it should be equally obvious that these metrics ignore conventional economic theory.

So one fact should be abundantly clear. We humans will have to deal with declining rates of oil production long before we bump up against the problems of complete depletion.

> In other words, oil production restraints
> will impact the price we can afford to pay for oil
> and availability of oil
> long before we deplete our oil reserves.

Calculating Available Reserves

But never-the-less, we have to start with the best estimate of available oil reserves that we can construct.

Two industry trade journals, Oil and Gas Journal[10] and World Oil[11], have published estimates of available liquid crude oil reserves by geographic region. These "Proven" reserves include "Conventional" crude oil and associated natural gas liquids. They exclude "unconventional" oil, refinery gains, reserve growth and undiscovered oil. In the following Table, we can tabulate the most optimistic estimate for each region. We then make two key corrections:

1. In working with the historical data, it became clear that the gradual depletion rate of North American reserves infers a larger base of available oil than current estimates. If we add 15.9 Bbl of oil to the North American base, the depletion rate makes more sense and appears to be consistent with industry reserve reporting practices.

2. As discussed above, the OPEC reserve data is specious. I did a net gain calculation, deducting the increased reserve claims made in the early 1980s and adding back my estimate of real oil discoveries. This calculation reduced Middle Eastern reserves by 258.3 Bbl. The lack of reliable OPEC reserve information presents a key problem

---

10 Oil & Gas Journal, is a weekly magazine of international petroleum news and technology, published by PennWell Corporation, 1700 West Loop South, Suite 1000, Houston TX 77027, http://ogj.pennnet.com/home.cfm

11 World Oil, PO Box 2608, Houston, Texas 77252, http://www.worldoil.com/, and World Oil Magazine, published by Gulf Publishing Company, provides monthly editorial content and handbooks on petroleum exploration, drilling and production.

for us in making our depletion calculations and is an area where we must do more work.

With these corrections, we are able to estimate "Proven" world oil reserves of conventional oil at 877.4 Bbl. assuming a sustained world price of at least $30.00 per barrel of oil.

Table 2
Proven Crude Oil Reserves (Bbl)

	Oil & Gas Journal	World Oil	Data Used	With Corrections
North America	49.9	50.9	50.9	66.8
C & S America	98.6	69.1	98.6	98.6
W. Europe	18.2	17.7	18.2	18.2
Asia Pacific	55.0	56.5	56.5	56.5
E. Europe & FSU	92.1	67.1	92.1	92.1
Middle East	686.3	662.5	686.3	428.0
Africa	117.2	94.9	117.2	117.2
Total	1,117.3	1,018.7	1,119.8	877.4
Source: Depletion Model				July, 2003

Given the OPEC correction, this estimate compares with the 2002 USGS estimate of 743.7 Bbl of "Known Reserves".

To complete our oil reserve estimates, we need to include "non conventional" oil and a reality check based on the oil we humans have produced to date. In the following Table, I have included estimated oil production prior to 12/31/2002, optimistic estimates of "non conventional" oil available in sands and shales, oil that is expected to be found in deep water or polar regions, and Natural Gas Liquids. Note: you will frequently see oil reserve quotations of 3.0 Tbl of oil. Be advised that these estimates usually include oil that has already been produced.

Table 3
Total Oil Reserves (Bbl)

Estimate November 2003		Remove Prior Production	Include Corrections
Production Prior to 12/2002	896.0		
"Proven" Crude Oil Reserves	1,119.8	1,119.8	877.4
Oil Sands and Shales	350.0	350.0	350.0
Deep water wells	60.0	60.0	60.0
Polar	30.0	30.0	30.0
Natural Gas Liquids (NGL)	300.0	300.0	300.0
Total	2,755.8	1,859.8	1,617.4
Source: Depletion Model, Campbell, et. al.			July, 2003

Now if you think what we have done is speculative, wait until you see what we have to do next. In order to identify all possible oil reserves, we must add "Undiscovered Reserves" and "Reserve Growth" to our calculations.

It is true. We humans keep on finding new deposits of oil. Although the rate of discovery is declining and the rate of discovery is not keeping up with oil consumption, it is still important to make an estimate of future oil discoveries. The USGS did this by looking at the geology of the rock that lies beneath each region of the world and then made a simulation of potential finds. Other oil industry participants have calculated the probable finds by extending known past production and discoveries into the future.

It is also true. Reserves "Grow". This is because when oil engineers first identify a new field, they make an estimate of potential reserves using various methods and some very sophisticated computer software. But ultimate probable production is an unknown until oil companies have drilled multiple holes into the ground. Once they have had a chance to look at drilled core samples and data on rock strata, they can make a better estimate of how much oil will be produced. Included in this calculation will be the potential additional reserve recovery that can be made by enhancing well production by injecting water, gas, etc. into the field. The resulting increase in production is called "Reserve Growth". It would appear that in a field of light oil, reserve growth may top 60 percent. Heavier oils, or oils in contaminated fields, flow more slowly. Reserve growth will be less. On average, it would appear that reserve growth will add 30 to 40 percent to the volume of the initial discovery. However, we must remember that some national oil companies like to use very optimistic estimates of their reserves and consequently their potential reserve growth is substantially less than 40 percent. In addition, oil company engineers are already using enhancement techniques in most of the older fields, so the potential for additional reserve growth is very limited. And finally, we are more likely to maximize production if there are no cultural impediments to the use of western technology. As a result of these caveats, future worldwide net reserve growth is more likely on the order of 12 percent over an extended period of time.

The following Table details how these two items increase the total estimated oil for each region. We start with "Proven" reserves and add both Undiscovered Reserves and Reserve Growth to calculate Estimated Total Reserves. Estimates are made on a region by region basis based on the age of existing fields, current production information and geological assessments. Assuming there are no barriers to exploration, production and transportation, we humans appear to have 2.175 Tbl of conventional and unconventional oil left on planet earth.

Table 4
Estimated Total Oil Reserves

World Oil Reserves at the end of 2002					
Recoverable crude oil (Bbl)					
	Proven Reserves	Un discovered Reserves	Reserve Growth	Total	%
Middle East	578.0	141.0	122.2	841.2	38.7%
W. Europe	18.2	3.9	5.2	27.3	1.3%
Central & S. America	253.6	31.2	28.7	313.5	14.4%
N. America	286.8	21.9	28.1	336.8	15.5%
E. Europe and FSU	222.1	39.7	30.2	292.0	13.4%
Africa	177.2	41.4	38.7	257.3	11.8%
Asia Pacific	81.5	12.6	13.4	107.5	4.9%
Total Estimated Oil	1,617.4	291.7	266.5	2,175.6	
% of Total Estimated Oil	74.3%	13.4%	12.3%		100.0%
Data from IEA, EIA/DOE, USGS, BP, Oil and Gas Journal, ASPO[12]					

On the assumption that a picture is worth a thousand words, this data is graphed in Figure 2, below. The Middle East is clearly a key player in the future of oil production. They have most of the

---

12 ASPO, The Association for the Study of Peak Oil & Gas

conventional oil (691.2 Bbl) and the most useful of the non conventional oil (150.0 Bbl). By contrast, some 49.4 percent of the total reserves in Central and South America are bound up in tar sands. For North America, 65.3 percent of total reserves are in the form of unconventional oil. These figures include 300 Bbl of NGL as well as 350 Bbl of oil to be derived from oil sands. Of particular concern is the fact that virtually all of the proposed oil sand and shale production is from two deposits - Western Canada and Venezuela, where gas resource and environmental factors will limit recovery operations. Most analysts separate conventional from non conventional oil in making reserve estimates because the infrastructure for transporting and refining conventional oil is in place.

The inclusion of unconventional oil (740 Bbl) in our estimates will be controversial, because if we are unable to exploit them as forecasted, then the following case studies are all much too optimistic. If we ignore unconventional oil, and focus our attention on conventional crude oil, then we humans only have 1.4 Tbl barrels of oil left (Table 5).

Figure 2
Estimated Total Oil Reserves by Region

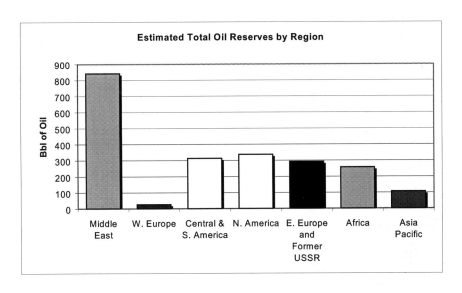

The largest oil users are located in the North American, Asia/Pacific and Western European regions. Combined, these consumer regions - which used 77 percent of the oil consumed in 2002, have only 16% of the conventional oil reserves.

Table 5
World: Conventional Oil Reserves By Region

Conventional Oil Reserves at the end of 2002					
Recoverable crude oil (Bbl)					
	Proven Reserves	Un discovered Reserves	Reserve Growth	Total	%
Middle East	428.0	141.0	122.2	691.2	48.2%
W. Europe	18.2	3.9	5.2	27.3	1.9%
Central & S. America	98.6	31.2	28.7	158.5	11.0%
N. America	66.8	21.9	28.1	116.8	8.1%
E. Europe and FSU	92.1	39.7	30.2	162.0	11.3%
Africa	117.2	41.4	38.7	197.3	13.7%
Asia Pacific	56.5	12.6	13.4	82.5	5.8%
Total Estimated Oil	877.4	291.7	266.5	1,435.6	
% of Total Estimated Oil					100.0%

Calculating Probable Production

But this series of estimates and calculations only tells us what oil may, or may not, be in the ground. To complete our exercise, we must know how much of this oil can actually be produced. For the purposes of this report, we are going to look at a period of 20 years - 2003 through 2022 so that we can compare the data thus developed with actual production and consumption data from the prior 20 years 1983 - 2002. We will refer as the future years - 2003 through 2022 - as the Forecast Period.

Net oil production must be evaluated on a region by region basis, taking into account the geographic, technical, production, transportation, financial, and cultural factors that will most likely impact production from 2003 through 2022. We need to:

- know the current rates of oil exploration,
- make an assessment of probable future discoveries,
- analyze current data on regional production,
- read the exploration and production opinions of geologists,
- ascertain how oil processing and transportation are likely to impact future production,
- assess the availability of capital,

42

- and determine how cultural influences will influence exploration and production contracts, oil field operations, oil processing facilities, and regional transportation.

Sorting through all the information, misinformation and outright deceit about regional exploration and production was the toughest and most time consuming part of the research that went into this report. Over 1,000 hours. I searched for clues, antidotal commentary and technical discussions that appear to have some level of merit. The greatest mystery surrounds the relatively closed social structures of the Middle East, and consequently that is where I was compelled to spend the majority of my time.

> One thing became very clear: we may have 2.175 Tbl of conventional and unconventional oil in the ground, but we can not possibly produce that much oil over the next 20 years.

In fact, it would appear that the maximum production is roughly 656 Bbl, about 30 percent of the available reserves. At the end of our Forecast Period - in 2022 - seventy percent of the available conventional and unconventional oil will still be left in the ground.

Why?

- Of the 2.175 Tbl of oil that we can expect to produce, only 66 percent - 1.435 Tbl - is conventional oil. Most of the oil we produce through the Forecast Period - about 92 percent - will be drawn down from these resources. By 2022, we will have extracted about 42 percent of this conventional oil. At that point, most of the oil left in the ground will be more difficult to extract (heavy oil, bitumen, sands and shales), of poor quality (chemical and brine contamination, etc.), and expensive to process.

- Unless we get really lucky, most of the conventional oil production will be from older - less productive - fields, or from newer fields that tend to be smaller in size (hence they deplete faster). Thousands of wells. But no flood of oil from any one field.

- New discoveries during the Forecast Period will take from 3 to 7 years to reach full production. That means their contribution to the world's demand for oil will be marginal because of the slow pace of oil field development.

- Deep sea exploration, while promising, will take time. It will be expensive. It will be very risky. It will be challenging to attract the needed capital.

By far the most daunting obstacles, however, are not technical, nor financial, nor a function of geology.

> Cultural conflict will be the primary barrier
> to oil exploration and production.

The Middle East has shown a growing inability to increase its oil production, preferring instead to rely on Saudi Arabia's excess capacity as a buffer against consumer demand. Cultural restraints work against bringing in enough technicians and engineers to upgrade the Saudi facilities to the full 30 - 35 Mbl per day production that would be needed to sustain world demand. In addition, there is a growing awareness of the region's economic dependence on oil revenue and this is stimulating concerns of conservation. It is understood that modest increases in oil production will be accompanied by greater consumer competition for available oil production, thereby driving a continuous increase in oil prices. Islamic militancy will also work to restrict production. Strikes, sabotage and political maneuvering promise to disrupt operations. Iraq remains a wild card. Its potential production will be determined by its political future.

Yet we need to get at least 31 percent of our oil from the Middle East in order to provide the world with the 656 barrels of oil in our projection.

Oil production in Western Europe is rapidly declining. Even the application of sophisticated oil recovery enhancement techniques will not stimulate greater production volumes. Western Europe will increasingly look to Russia, the Middle East and unconventional resources (polar and deep sea exploration) for its oil requirements. Western Europe will provide only 3 percent of world production.

There is more oil to be found in Central and South America. The Orinoco tar sands promise to yield a steady - if unspectacular - increase in oil production. Off shore deep sea exploration will increase potential production. Yet we can expect environmental, cultural, capital and technology constraints to limit production to about 13 percent of the total oil produced between 2003 and 2022. Labor problems in Venezuela, for example, have already disrupted oil production and they are likely to erupt again.

There is also more oil to be found in Africa . Off shore deep sea exploration is certain to increase potential production. Yet we can expect cultural, capital and technology constraints to limit production to about 18 percent of the total oil produced between 2003 and 2022. As the Nigerian experience has shown, some of the cultural constraints will be less about oil then they are about tribal conflict and political power struggles.

Although China, Japan and Australia (among other nations) are actively looking for new oil deposits in the Asia Pacific region, the prospects are not promising. According to the USGS, the geology of the region works against finding any "blockbuster" fields. This region will only contribute 8 percent of the world's oil production through the Forecast Period.

With the collapse of the former Soviet Union, oil production in the Eastern Europe and FSU region declined rapidly. The political situation, however, has stabilized, and oil companies with sufficient financial strength have emerged to reopen regional exploration and production. Despite optimistic claims, however, actual production is expected to be about 13 percent of world consumption through the Forecast Period.

Which leaves North America (Mexico, U.S.A., and Canada). Environmental constraints will limit oil production in the Gulf of Mexico, off the coasts of Florida, California and New England, and in the polar regions until later in the Forecast period. Exploration will be restricted until an oil crisis occurs. By then, however, it will be too late to contribute to the needs of consumers during the Forecast Period. Even with the (somewhat controversial) production of oil from the oil sands of Western Canada, this region will only contribute about 13 percent of the world's oil between 2003 and 2022.

Actual oil production could be more (or less) than shown in the following Table and Chart..

Table 6
Estimated Maximum Oil Production 2003 - 2022

Oil Reserves at the end of 2002				
Recoverable conventional and non conventional crude oil (Bbl)				
	Maximum Reserves	Maximum Production	% of Region's Reserves	% of World Production
Middle East	841.2	205.3	24.4%	31.3%
W. Europe	27.3	19.8	72.5%	3.0%
Central & S. America	313.5	86.9	27.7%	13.2%
N. America	336.8	87.9	26.1%	13.5%
E. Europe and FSU	292.0	88.2	29.5%	13.1%
Africa	257.3	119.0	46.2%	18.1%
Asia Pacific	107.5	50.9	47.3%	7.8%
Total Estimated Oil	2,175.6	656.0		
% of Total Estimated Oil				100%

Figure 3
Maximum Oil Production by Region
2003 - 2022

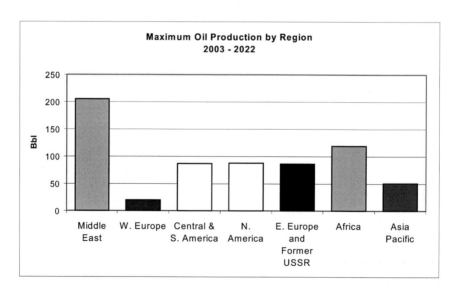

46

The Oil Depletion Model

Assuming that no one rolls over a rock and finds another 1 Tbl of oil (highly unlikely), we have established a reasonable estimate of how much oil will be available to the world from 2003 through 2022. Over a 20 year period, we will produce - and then consume - up to 656 Bbl of oil. The next step in our quest to answer the three fundamental questions raised by this study is too build a model that permits us to propose, and then test, alternative production scenarios.

The model we will use is a classic double entry bookkeeping spreadsheet.

On the left hand side of our spread sheet we estimate average daily oil production for each year by four producer regions: North America (the United States, Canada and Mexico); the Middle East (primarily OAPEC[13]); EurAsia (Western Europe, Eastern Europe, the nations of the Former Soviet Union or FSU, and the Asia Pacific region, including China and India); and Rest of World (including Africa, Central America and South America). We include data for all forms of oil (conventional and non-conventional oil). For each cell in the column of production we calculate the annual rate of change in production and the percentage of available reserve depletion. We calculate total average daily production by year by adding together the four columns of production with an additional column that estimates net refinery gains. The total average daily production is multiplied by 365 to get annual oil production. In 2002, our average daily production was 73,936,000 barrels of oil. Total Annual production, including the addition of refinery gains, was 27,423,910,000 barrels of oil.

On the right hand side of our spread sheet is oil consumption by four geographic regions: North America (again including Mexico, the USA and Canada), Western Europe (where Germany, the United Kingdom, France and Italy are big consumers of oil), Asia/Pacific (primarily China and Japan), and Rest of World. It is very important to understand that - with the exception of North America - the geographic configuration of the regions that produce oil are different from the regions that consume oil.

There is a consumption column for each region along with a column for the annual rate of change in consumption and the annual net change in consumption. Total annual average world consumption per day is multiplied by 365 to get total annual consumption and there is a column that calculates the annual rate of change of world consumption.

In the center of the spreadsheet is a column that nets annual production versus consumption and a column that averages this net

---

13 OAPEC - nations that belong to the Organization of Arab Petroleum Exporting Countries

data for each succeeding 2 year period. Historically, production - on average - has always exceeded consumption and while there may be a gross mismatch for a single year, the two year average tends to be less volatile. Thus, in making our production vs. consumption estimates, we have assumed that average two year production must exceed average two year consumption. It is rare for consumption to exceed production for more than 3 years in a row.

Please note that we are counting barrels of oil that are actually consumed for gasoline, heating oil, chemicals, cosmetics and so on. We are NOT determining the demand for oil which tabulates how much oil would be consumed if it were readily available at an acceptable price. As we have seen in the 1970s, the demand for oil can be far greater than the actual consumption of oil, because oil shortages force a reduction of consumption. Therefore, in our model, if projected production shortfalls force a reduction of world consumption, the average shortfall for each succeeding two year period has typically been constrained to less than 200 Bbl by forcing a reduction of consumption according to the economic activity of each region. Conversely, if world production exceeds consumption, it has been assumed that regional economic activity will gradually permit increased oil consumption. Decreases or increases in consumption versus demand by region have been done by calculating the economic impact of a shortfall or excess on each region. Long term consumption patterns have been aligned with regional economic activity.

For three scenarios we have added columns to our model that calculate the impact of the annual change in consumption on annual Gross Domestic Product, the Rate of Inflation, the Rate of Unemployment and the price for a gallon of gas in the United States. Other columns tabulate the average annual price for a barrel of oil, the annual rate of change in the price for a barrel of oil, the natural growth in oil demand for the United States and the world (assuming no constraints), unsatisfied world demand, unsatisfied world demand adjusted for reduced consumption, and a gasoline price inflator (to compensate for the increasing cost of refining oil as the quality of the oil stock declines).

Each scenario has been constructed with due attention to the economic and cultural constraints discussed in this report. But in the final analysis, the oil depletion model does not attempt to nail down a guaranteed view of the future. It can not be done. Instead, it is a "what if" model. A working hypothesis. It lets us develop economic impact scenarios based on alternative sets of assumptions. These assumptions drive the changes, results and values shown for each data series. Change the assumptions and you change the results. We need to improve the data used and the calculation methodology as we identify quantifiable improvements.

48

It should also be noted that it is possible to use this model as a means of calculating these data series for other nations or regions. Of particular interest would be the primary consuming nations of Western Europe, along with China and Japan.

Economics 101

Consumption and GDP

When we buy goods and services, we are engaged in an act that will lead to their consumption. We may use (consume) them immediately (as with goods such as gasoline or services such as haircuts, etc.), sometime in the future (as we typically do with canned food, clothing, etc.), or over a long period of time (refrigerators, automobiles, etc.). We use Gross Domestic Product (GDP) as a way of measuring the dollar value of everything an individual nation, a geographic region, or the world is able to produce within a given time frame (a month, a quarter or a year).

As one may suspect, there is a relationship between consumption and GDP. As consumption rises, there is an attendant increase in the demand for goods and services that results in greater production (and hence GDP). Conversely, when consumption declines, so does GDP.

Historically, there has also been a relationship between oil consumption and GDP. In the past, the increase or decrease in GDP (which measures the production of goods and services), tended to drive the demand and consumption of oil. The more goods and services we produced, the more oil we needed in order to produce our goods and services. We used more gasoline to move things and people, we used more oil for the generation of electricity, and we used more oil as a feedstock for the production of goods (plastics, chemicals, cosmetics, drugs, and so on). If on the other hand, the consumption of goods and services declined, then GDP and oil consumption also declined.

In developing the economic impact analysis for the oil crisis scenarios described in this report, estimates of GDP are tied to estimated oil consumption and estimated oil pricing. In so doing, our formulae must account for the fact that the quantity of oil used per unit of GDP has been changing. While the mature economies of the world are becoming more efficient in their use of oil (using less per unit of GDP), emerging economies have tended to use more oil per unit of GDP. We also need to include in our formulae the concept that sharp increases in the price of oil will force people to consume less oil (almost immediately because we cannot afford to pay a higher price) and sharp decreases in the price of oil will stimulate greater consumption (although this takes a longer time because lower consumption has usually been associated with recessive economic conditions that take time to improve).

There is another problem. In the past, GDP and oil consumption have tended to move in tandem (more or less - oil consumption tends to be more volatile than GDP). Changes in GDP drove changes in oil consumption. But as we move from a world economy that has enough oil to meet demand, to an economy that must deal with periodic oil shortages, then the reverse will be true. *Oil shortages (or availability), along with the price of oil, will tend to drive the growth or decline of GDP.* In addition, the price of oil will rise until there is a balance between supply and demand. But this relationship will also be more complicated than it has been in the past. There is a high probability that future oil markets will be characterized by arbitrary oil prices. It will take longer for the supply versus demand mechanism to resolve any imbalances. In addition, oil consumption for transportation will evolve from an emphasis on individual vehicles (my car) to mass transportation (including ride sharing), moderating the normal impact that the supply versus demand mechanism would have on pricing.

In determining how changes in oil production (availability) will impact the price of oil, we must consider whether or not the changes in the price of oil are based on a willing buyer and a willing seller in a market that is free to move according to negotiated supply and demand pricing; we must factor in the impact of other inflationary forces; we must include the length of time that these changes take to occur; and we must determine the status of the economy at the time these price changes occur. And finally, the price of oil and GDP tend to have an *inverse* relationship.

Confused? Just remember.

We are moving from a world economy that enjoyed excess oil capacity
to a world economy dominated by
chronic, severe, and highly volatile shortages.
The GDP of all nations will have a volatile response to these shortages.

Rate of Inflation

It's safe to say that increased oil prices will drive up the Rate of Inflation. Although the price of oil tends to be more volatile than the Rate of Inflation, there is a correlation. Rates will be highly volatile as periods of oil shortage alternate with months of surplus. If the price of oil were the only driver of inflation, then inflation would skyrocket. But there are other factors that must enter into our calculation. The combination of higher prices and sporadic shortages will drive an increase in unemployment, restrict consumption and disrupt both the production and distribution of goods and services. Productivity will decrease. Lower interest rates will only marginally help the economy because oil shortages will disrupt the flow and use of money in the economy. These impacts are all *deflationary*. Thus in our formulae for calculating Inflation, we must offset the inflationary impact of higher oil prices with the recessive impact that oil prices and shortages will have on the economy.

We also have to include the deflationary impact of unemployment on regional demand and GDP. Over the last two decades, over 60 percent of displaced white collar workers found new jobs that paid less than they were making before becoming unemployed[14]. For white or blue collar workers living in the highly developed economies of the industrialized world, either the Political or the Production Crisis discussed in this report will exacerbate this problem. Lower pay means lower oil consumption and a declining GDP.

In the economic impact analysis used for the Best Case, Production and Political scenarios, the Rate of Inflation has been tied to the rate of change in the world price for oil as well as a calculation of unsatisfied oil demand. It works this way. The demand for a scarce commodity will drive up its price. As the price of oil goes up, changes in consumer spending choices gradually reduce real demand. This in turn reduces the upward price pressure on the commodity. As long as real demand (how much oil we would consume if it were readily available) exceeds actual consumption, the difference is called unsatisfied demand. All three scenarios reflect greater volatility in the Rate of Inflation because we are moving from a world economy that enjoyed excess oil capacity to a world economy dominated by chronic, severe, and highly volatile shortages. This volatility will drive corresponding changes in unsatisfied demand and inflation as consumers adjust to shortages by bidding up the price of oil based products.

Unemployment

Any oil crisis will drive up the rate of unemployment. Primary factors include: a decrease in consumption of goods and services, the horrific disruption of transportation and a fear driven decrease in capital spending. In the Best Case scenario described in this report, oil shortages create a mildly recessive condition in the economy. The Production Crisis drives us into a genuine - and long term - recession. The Political Crisis scenario describes an economy that plunges into a depression.

Future estimates of unemployment must include a consideration for persistent oil shortages and the resulting volatility of oil prices. The annual change in oil consumption is therefore a better guide to

---

14 McKinsey Global Institute, reported in Business Week, December 8, 2003, pp. 71.

estimated unemployment than the price of oil. We can assume that in periods of restricted supply, nations will consume all the oil they can get up to the point where there is sufficient oil to sustain current economic activity. The level of economic activity will be directly proportional to available oil supplies. Oil consumption and unemployment have an *inverse* relationship. As oil consumption increases, unemployment will decrease - and vice versa.

For example, if a nation has sufficient free cash flow, consumer demand, and non-oil resources to increase its GDP by 1.3 percent for a given year, then its oil consumption also needs to increase by 1.3 percent (ignoring the impact of changes in energy efficiency). If there is a surplus of available oil, then a growth rate of 1.3 percent is achievable. However, if there isn't enough available oil to permit the potential increase in consumption, then economic activity must grow at a slower rate. If the shortage is severe enough, economic activity will be forced to decrease.

I relied on an inspection of how unemployment has acted in previous recessions (and the depression of 1929) in order to make an educated guess of the projected rate of unemployment that will occur in an oil crisis. Hopefully, there is a really good econometric model on a big computer somewhere that can be used by a really smart economist to improve on the results of my mental calculations.

Global Impact

Although this report only deals with the economic impact of an oil crisis on the United States, these calculations could be duplicated for every nation on this planet.

> Any oil crisis will have a global reach,
> sparing no nation from its pain and hardship.
> The industrial nations of North America, Europe and the Pacific Rim
> will be hit the hardest
> because they have the most energy intensive economies.

In making assessments of global oil consumption, we have to factor in the rapid economic growth of nations such as China and India, increasing demand in third world countries, recognize the interaction of regional economies (consumption in America creates jobs in China, and so on), and make some assumptions about the development of alternative forms of energy that will eventually reduce the demand for oil. The oil crisis described in the Best Case and Production scenarios may also produce a panic in world financial markets. The Political Crisis will definitely cause these markets to collapse.

There is one other factor that we must consider when we make an estimate of how an oil crisis will impact global production and consumption.

People.

In making the production assumptions, it can be assumed that as the price of oil increases, limited additional production will come on-line to satisfy demand. Thus, if Middle Eastern producers restrict production, the resulting shortages will drive up the price of oil and this in turn will stimulate additional production in the Pacific Rim, North America, EurAsia, Africa and South America. This has been the traditional economic response to shortages. But we must modify our production assumptions based on social responses as well as the limitations of elasticity discussed above. Environmental concerns will act as a drag on new production, exacerbating oil shortages and prolonging the recessive impact of an oil crisis. Islamist influence will have a negative impact on production and transportation in the Caspian, North African, West African, and Pacific Rim oil fields. The transition to alternative fuels is also both a technical and a cultural challenge. And of course, if our cultural problems can not be constrained, regional or world war is always a possibility.

1929

The last comparable economic shock to the world economy occurred in 1929. Severe deflation dropped the American Consumer Price Index (CPI) by over 23 percent to a low of 13 in 1933. The years 1934 through 1940 were characterized by modest changes to the CPI. Unemployment increased by 728 percent, from 1.55 million in 1929 to 12.83 million in 1933. America did not reach a full employment economy until 1942 - 13 years after the collapse of the economy in 1929. American Gross National Product (GNP) plummeted 9.4 percent in 1930, 8.5 percent in 1931, 13.4 percent in 1932, and 2.1 percent in 1933. It bounced back from 1934 through 1937, was negative again in 1938, and then increased through the years of WW2.

By 1932, industrial stocks had lost 80 percent of their value, 40 percent of the banks had failed, and international trade had fallen by more than 60 percent.

If a Political Crisis occurs, the world will suffer the same kind of devastating economic volatility that it did in 1929. If oil production simply fails to meet consumer demand over a long period of time (there is no political crisis), then the Best Case and Production scenarios become more likely. In all three cases, the economic trends will be irreversible unless we humans develop a suitable alternative energy system.

Chapter 4   LOW PROBABILITY

We want to believe we can pump oil forever. We want to ignore the fact that oil is a finite resource.

Introduction

Seven alternative scenarios were developed for this report. Each one attempted to characterize the impact of a unique set of events and assumptions on oil production and consumption for a 20 year period from 2003 through 2022. We used 20 years of historical data (1983 through 2002) as a reference point for each regional production and consumption forecast. Where available, data going back to 1970 was used as a means of calibrating the integrity of future estimates.

There are - as should be obvious - a number of alternative oil depletion scenarios. The oil depletion and impact model permits us to explore them and to compare the results. Scenarios are not predictions. Rather, they permit us to make and challenge assumptions, encourage debate, and profile the probable result of each alternative hypothesis. Scenarios are tools that give our evaluations focus, permit us to deal with the unexpected, and characterize the results of dynamic circumstances. And that is the value of the model. It encourages us to ask "what if" questions and then to model the probable behavior of oil production and consumption based on a unique set of assumptions[15].

In the interest of brevity, I selected the four most probable scenarios for presentation in this report. It should be noted that these four scenarios - assuming no catastrophic events occur - probably cover the range of possible outcomes. Additional changes to our data will most likely mimic one of these four scenarios. The result will be worse - if production goes down. The result will be better - if production increases. The reason?

The oil market has transitioned from a consumer driven market to a producer controlled market.
Future consumption will be limited by production.

The No Change and Best Case scenarios presented in this chapter describe an optimistic view of oil exploration and production. Everything works. Everybody cooperates. For the 20 year period, 2003 - 2022, oil depletion is a relatively minor problem.

---

15 For additional information on oil production, consumption and depletion, see Appendix 1.

The "No Change" Case

The No Change case assumes that the next 20 years are going to be just like the last 20 years. This is the case for those who believe (or want to believe) there is no problem. There is ample oil and we can find more. Production of oil from existing fields, new wells, oil shales and sands, and deep water and polar oil fields will come on-line in time to satisfy a growing demand. There are no technical or cultural impediments to restrict production.

The assumptions:

- North American oil production peaked in 1970. Although it briefly recovered in 1985, it has been gradually decreasing ever since. By 2002, it had declined to 14,163,000 bl per day. In making a forward estimate for the 20 year period between 2003 and 2022, we have assumed an average Compound Annual Rate of Decline (CARD) of minus .2 percent. This the same rate of decline that North American oil production experienced during the 20 year period between 1983 and 2002.

- We have assumed that for the next 20 years Middle Eastern oil production continues to increase at a Compound Annual Growth Rate (CAGR) of 3.1 percent, the same rate of increase it was able to achieve between 1983 and 2002.

- EurAsian oil production peaked in 1987 at 17,240,000 bl per day. Oil production in 2002 was approximately the same as it was in 1983. Hence, the CAGR for oil production is zero. We have applied this same CAGR to the next 20 year period.

- Rest of World (ROW) oil production has been estimated as having a CAGR of 2.7 percent, the same rate of growth experienced in the prior 20 year period.

- Refinery liquids increase daily production by an average of 1.62 percent.

- Total annual oil production increases from 28.1 Bbl in 2003 to almost 39.9 Bbl in 2022, yielding a CAGR of 1.89 percent.

- North American consumption continues to increase by a CAGR of 1.4 percent.

- Western European consumption continues to increase by a CAGR of .9 percent.

- Consumption in the Asia pacific region continues its robust CAGR of 4.0 percent.

- ROW consumption continues to decline at a CARD of minus .3 percent.

The results: The future can not possibly replicate the past.

For one thing, North American oil production is declining much faster than the underlying assumptions used in this Scenario would suggest. In order for the future to replicate the past, North American oil producers would have to find <u>and produce</u> an additional 13.3 Bbl of oil that currently does not exist. Given the oil production parameters discussed in this report, this is highly unlikely[16].

It is also highly unlikely that the Middle Eastern producers will continue to increase oil production at a CAGR of 3.1 percent. In order to do this, Middle Eastern governments would have to dramatically increase the number of western technicians and engineers that are working in their oil fields, make substantial additions to their oil production infrastructure, minimize bureaucratic mismanagement, and find multiple billions of dollars of additional investment. We would also have to assume a stable cultural and political environment as well as the continuing control of Islamist anti-western sentiment. It is doubtful that these nations have the necessary political control and motivation to accomplish these tasks.

With the conversion of the Russian oil companies to a competitive business model, their ability to find and produce oil has been sharply increased. Nations in the EurAsia region have shown a more assertive attitude in the development of oil bearing properties. During the 20 year Forecast Period covered by this report, it is reasonable to predict that the EurAsian region should be able to increase its oil production by 34.1 Bbl over the production assumptions used in this Scenario.

On the other hand, existing exploration and production trends - if they continue - will reduce the growth of oil production in the ROW regions by 14.7 Bbl below the volumes that would have to be assumed for this Scenario.

Regional oil consumption trends have been changing and will continue to change. Assuming there are no impediments to production or consumption, and assuming oil prices are non-inflationary, then natural demand, real demand and consumption will be equal. Average annual consumption will therefore be driven by the economic growth of each region. Based on economic projections for each region, the average annual consumption for our forecast period, 2003 - 2022, will decrease from a Compound Annual Growth Rate (CAGR) of 1.4 to 1.3 percent in the North American region, from .9 to .7 percent in Western Europe and from 4.0 to 3.6 percent in the Asia Pacific region. Because of continuing economic development, Rest of World (ROW) demand will increase from a CARD of minus .3 percent to a CAGR of plus .5 percent. Over this 20 year period, the CAGR of world oil consumption is predicted to increase from 1.4 to 1.8 percent.

---

16 Remember that our definition of North America includes Canada, Mexico and the United States.

Conclusion: The regional patterns of oil production and consumption are changing, and will continue to change, making it impossible to assume that the oil market for the next 20 years will be just like the market has been for the last 20 years. Of particular interest is the prospective limitations on Middle Eastern production, some gains in EurAsian production (mostly in the Former Soviet Union), and new fields off the coast of Africa, South America and the Arctic circle. China will become the number two oil importing nation. Oil consumption in third world nations will increase.

Because this scenario ignores the realities of changing patterns in both the production and consumption of oil, the probability that this scenario presents an accurate projection of the future is less than 10 percent.

The Best Case Scenario

The Best Case Scenario imposes a more realistic assessment of probable production and consumption on the 20 year period from 2003 to 2022. Historically, regional oil production has been erratic and this characteristic plays an important role in the availability of oil. Production shortfalls must be accompanied by reduced consumption. Although we assume that western military presence will protect a relatively modest increase in Middle East production, this scenario primarily relies on geographic, technical and economic factors to explain changes in oil production.

The Best Case Scenario also corrects the technical and production assumptions of the No Change case. We assume that over the 20 year Forecast Period all of the producer regions reach 100% of their maximum available production. It also assumes oil consumption for each consumer region responds to changes in the availability of oil (shortage or surplus) as projected, and that oil exploration, production, transportation, refining and distribution are not subject to any cultural disruptions.

In doing the Best Case Scenario, we attempt to replicate - with suitable adjustments for projected regional production - the historical pattern of oil production. Each producer region is examined separately. On a regional basis, production has been erratic and there is no reason to believe that this will change. World oil shortages and surpluses are factored into a calculation of opportunity demand.

In making our examination of user region oil consumption, we must adhere to the rules of economics. Consumption can not exceed production - on average - for more than 2 years. There simply is not enough buffer to sustain consumption if there are deep cuts in production. Furthermore, the availability of oil - or oil shortages - not only impact consumption, they drive the economic health of consumer nations. This is an interactive scenario. Decreased production forces reduced consumption. Excess production permits consumption to rise over time. This is, after all, a typical producer controlled market profile.

Assuming that environmentally sensitive areas remain off limits to exploration and production, North American oil production declines by a projected Compound Annual Rate of Decline (CARD) of minus 2.0 percent. Although total annual production for this 20 year period exceeds 87.9 Bbl, by 2022 average daily production has decreased to less than 9.6 Mbl per day and North America has produced 26.1 percent of all the oil (including deep water, polar and oil sands) that will ever be available in this region. Worse, it will have produced over 70 percent of its remaining conventional crude oil. North American oil production is more vigorous in the early years of the forecast period from 2002 through 2022. This is reality. North American oil companies are currently producing every drop of economically profitable oil they can.

But the net impact is to reduce the amount of oil that can be produced in the out years of our forecast period.

Figure 4 shows the long term trend of North American oil production by calibrating the Rate of Change in annual production for each succeeding year. In the No Change Case we made the statistically nice assumption that oil production will increase or decrease each year by approximately the same percentage. It is an orderly Scenario. Unfortunately, it is not reality. Regional oil production tends to be erratic with annual fluctuations that occasionally exceed 5 percent. The Best Case Scenario, replicates this erratic behavior by building a series of estimated annual gains or losses into the annual oil production estimates for each region. The North American profile, below, shows the results of this simulation. Using data from 1970 through 2002 for purposes of comparison, we can replicate a pattern of oscillation and apply it to the Forecast Period. By 2022, production has dropped by 33 percent from the 14.2 Mbl produced per day in 2002..

Figure 4
Rate of Change, North American Oil Production
Best Case Scenario

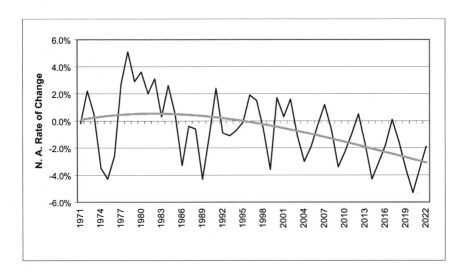

Given the assumption of no compelling cultural constraints, Middle Eastern oil production increases by a CAGR of 2.1 percent, from 21.8 Mbl per day in 2003 to 32.1 Mbl per day in 2022. We have assumed that a portion of this increased production will be the result of using up all of Saudi Arabia's existing production buffer. By 2022, the world will no longer have any way to protect itself from erratic changes in regional production. EurAsian and ROW nations are also working hard to produce more oil. With a CAGR of 2.2 percent, EurAsian production reaches 25.6 Mbl per day by 2022. ROW production increases at a CAGR of 2.0 percent to 33.5 Mbl per day in 2022.

The Best Case Scenario assumes that we humans will have used up 30.2 percent of all the conventional and unconventional oil left on this planet by 2022. We will have used up 42 percent of the conventional crude oil. Total world oil production increases at a CAGR of 1.5 percent from 28.0 Bbl in 2003 to 37.4 Bbl in 2022. In contrast to some oil depletion models that show a bell shaped curve with steep slopes, the Best Case and Production Crisis scenarios used for this report both yielded essentially flat production for several years at the peak. In the Best Case scenario, world oil production stalls at 36 to 38 Bbl per year from 2018 through 2026, the peak year of production. Look at the 4.8 percent drop in production shown by Figure 5 for 1975. Remember how the associated oil shortages caused long lines at the gas station? Another 15 percent reduction of production occurred from 1980 - 1983. Since then, total production has grown at a reasonably even pace, even through individual regions have experienced erratic fluctuations. Why? Because Saudi Arabia had excess capacity. If one of the other regions had a temporary shortfall in oil production, Saudi Arabia opened the spigot to make up the difference.[17]

Although world production is still increasing, the rate of change has flattened by 2022, signaling an impending down turn in production. It should be noted that if we only consider conventional crude oil reserves (and therefore ignore unconventional oil production), a Rate of Change calculation would appear to indicate that world oil production peaked in 2000. In this scenario, however, peak oil production is delayed until 2026 by a growing reliance on the production of unconventional oil, including oil equivalents derived from Natural Gas Liquids, shales, polar and deep water resources.

---

[17] Figure 5 includes historical data from 1970 - 2003.

Figure 5
World Oil production: Best Case Scenario
1970 - 2022

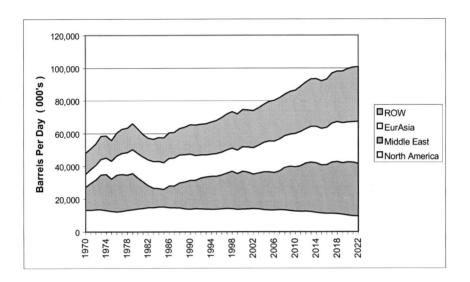

For the purposes of this Scenario, it has been assumed that North American natural demand will grow at a CAGR of 1.3% and Western European natural demand will grow at a CAGR of .7 percent. Short term shortages restrict Asia/Pacific and ROW consumption to a CAGR of 2.8 percent and .3 percent respectively.

The results: **This optimistic assessment could become our future**.

But probably not.

Although consumption matches production, and world oil consumption actually increases from a 1.4 percent CAGR in the 1983 - 2002 timeframe, to a 1.5 percent CAGR during the Forecast period, the loss of the Saudi Arabian production buffer ties world oil consumption ever more tightly to available production. Temporary production shortages do occur, and when they do - they force a reduction of world oil consumption.

In order for production to equal consumer demand (not consumption), world oil production would have to increase by a CAGR of 1.8 percent. It will not. As shown in Figure 6, by 2014, the declining rate of change in world oil production will force a decline in consumption. For purposes of comparison and information, Figure 3 includes historical data from 1970 - 2003.

> Worldwide consumer demand exceeds available oil for consumption in 8 out of the 10 years, 2013 - 2022.

The cumulative failure of world production to satisfy world demand drives up the price of oil to $66 per barrel by 2022.

Figure 6
World Annual Change in Production
Vs. Average World Price per Barrel of Oil
Best Case Scenario

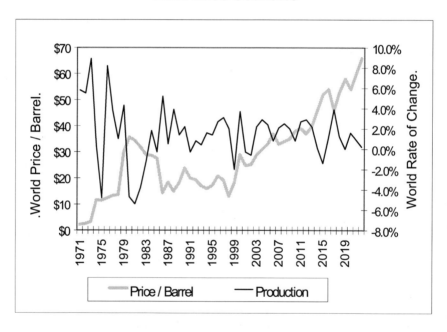

We must anticipate that the technical disruptions of consumption shown in Figure 7 will occur during the Forecast Period. Sure. The timing and the duration will be different from the assumptions used to create Figure 7, but they will happen. Continuing shortfalls will force a gradual decrease in the trend of North American oil consumption. Figure 7 includes historical data from 1983 - 2003. Note the significant volatility in the rate of change for North American oil consumption.

What does this "Best Case" scenario show?

From 2013 - 2022, North America will not have enough oil
to meet natural demand in 5 of these 10 years.

By 2018, economic growth has been curtailed forever. North
American consumption stalls at less than 31 Mbl per day.

Figure 7
Rate of Change
North American Oil Consumption
Best Case Scenario

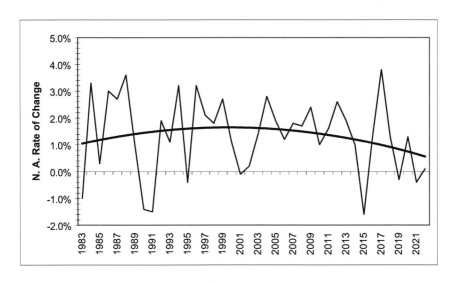

In a producer controlled market, oil consumption must track oil
production because consumer demand will be moderated by the
availability of product. We also know that oil consumption and
economic activity generally move in the same direction. The erratic
nature of oil production may begin to curtail North American
consumption, and hence economic activity, by 2006. It will most
certainly impede economic activity by 2013. All consumer regions -
including Western Europe, China and Japan - will experience a similar
impact. In some years, oil will be plentiful. In other years, there will be
shortages.

There is, of course, a relationship between oil consumption and GDP. In Figure 8, we are able to illustrate this relationship for U.S.A. Gross Domestic Product versus North American oil consumption going back to 1981. The rate of change for U. S. GDP has been declining - even through the economic boom of the late 1990s - and there is no reason to believe that oil shortages will help improve this trend. The formulae used to replicate history appears to show a distinct decrease in the GDP of the United States beginning in 2012 - 2014.

It should be noted that one could draw a similar consumption versus GDP chart for every industrialized nation on this planet. Or for the world as a whole. Unless there is a radical upward change in the projected availability of oil, World GDP has only one way to go - down.

Figure 8
Rate of Change:
N. A. Oil Consumption & U. S. A. GDP
Best Case Scenario

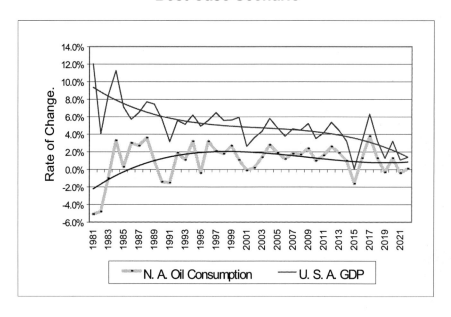

If oil consumption is forced into a decline, or fails to grow fast enough to meet natural demand, then there must be a corresponding negative impact on unemployment. If oil consumption is restricted by available production, then GDP will decline. Under these circumstances, GDP and unemployment will tend to have an inverse relationship - a declining GDP will be accompanied by increasing unemployment. Unemployment increases because any industry tied to the consumption of oil - including discrete manufacturing, transportation, retail trade, etc. - will suffer. In Figure 9, we track U. S. A. GDP and Unemployment trends from 1970 - 2002 using historical data, and projected trends for this scenario from 2003 - 2022.

Look at the trends in Figure 9. By 2022, unemployment in the U. S. A. is projected to exceed 7 percent - forever.

Figure 9
U. S. A. Unemployment and GDP
Best Case Scenario

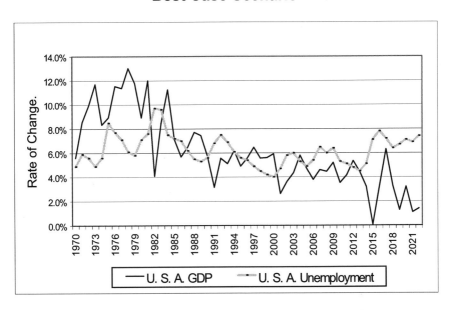

Conclusion: If this optimistic scenario is the best we can do, then our future does not look very encouraging. The Best Case scenario shows that we are entering a new era. Consumption will be tied ever more closely to the risks of available production. Even relatively benign disruptions to oil production will drive economic recession.

And the bad news. In order for this optimistic scenario to work, the oil producer nations would have to be willing - and able - to make substantial increases to their average annual production. They would have to be ready, willing and able to match consumer demand, without regard to their potential pricing power in a producer controlled market. Avoiding larger shortages than shown in this scenario would require an effective spirit of cooperation between producer and consumer nations. Everything has to work. No political posturing. No bureaucratic mistakes. No under the table deals for extra oil. No terrorist disruption. No cultural or labor conflict.

Do you believe that is possible?

The probability that the Best Case scenario presents an accurate projection of the future is less than 40 percent.

Chapter 5 Islamist Challenge

We wish there were no war in Iraq. We wish the poor people of producer nations could benefit from their nation's oil wealth. We wish Islamist Jihad did not exist. We wish there were peace in the Middle East. Our wishes are popular and politically correct. Reality is not.

Prelude

American interest in the Middle East increased after World War One. In a sense, the United States followed the British, who had commercial interests in Egypt, Iraq, Iran, and Palestine; the French whose influence was primarily concentrated in the Maghrib of N. Africa, Syria, Lebanon, Egypt and Iraq; and Italy with its ties to Libya, Albania and Ethiopia. Europeans controlled trade, finance, industry and land use in many of the countries of this region. In general, the Europeans believed they had a superior civilization, the right to govern, and a mission to bring civilization to the Middle East. During the period from World War One to World War Two, Anglo-American oil companies established a firm control over Middle Eastern oil production. They furnished the engineering, management and capital for exploration and production. The business was run to suite their own interests and they had a great deal of control over the world price for oil. Because it had a growing oil industry of its own, America was relatively independent of Middle Eastern oil production.

Selected tribal leaders became prominent members of the elite European group that ruled each country. Their children frequently went to European missionary schools, and some went on to attend European colleges and Universities. There was an influx of Jewish migration into Palestine from 1918 to 1939, prompting a growing feud that produced an armed clash in 1936. Given the economic and cultural disparity between them, it should not be surprising that conflict emerged between European and Arab. Resentment occasionally erupted in armed clashes. There was also a fundamental economic problem. The mechanization of farming displaced millions of workers who were forced to migrate to the cities of the Middle East for refuge. Unfortunately, unlike Western Europe, the emerging Arabian nations did not have a burgeoning manufacturing base to absorb these workers. Poverty generated dissent, eventually forcing Great Britain to grant King Faysal of Iraq some level of independence in 1930, and to reach an accord with Egypt's Wafd Party in 1936. 'Abdal-'Aziz ibn Sa'ud used the period from 1914 to 1939 to consolidate his power over the Arabian peninsula.

Economic devastation in Europe, a growing aversion to colonialism, and increased Arab nationalism forced a reduction of European influence in the Middle East and North Africa after World War Two. The League of Arab States, founded in 1945, unified a common cause - support the Palestinians against the Jews. The Suez Crisis of 1956, the Algerian war of 1954 - 62, the Yom Kippur war of 1973, and the Iranian revolution of 1979 confirmed a growing distrust and contempt for the west. There was a revival of fundamental Islamist tradition.

During the 1950s, world demand increased rapidly as oil gradually displaced coal as a primary source of energy and consumers developed a preference for personal transportation. Big Oil was encouraged to increase its stake in the Middle East. It was recognized that these oil fields held a substantial percentage of the world's reserves. By 1960, the Middle East and the Maghrib produced approximately 25 percent of the world's oil. It was cheaper to explore, produce and ship oil from the Middle East than it was from the oil fields of the west.

Following World War 2, American politicians and administration officials began to forge close ties to the Royal Family of Saudi Arabia. Oil was the end-game that compelled political and economic accommodation. American oil production peaked in 1970. From then on, its growing demand for oil would have to be satisfied through the development of international petroleum resources.

American foreign policy in the region had three goals during the cold war: keep the Russians out, support Israel, and control the flow of oil. The United States emerged as a power broker in the Middle East, drawing Egypt, Morocco, Tunisia, Jordan and Saudi Arabia into closer ties. Unfortunately, Israel's victory over Egypt, Syria and Jordan in 1967 served to unify a growing hatred of Israel and its primary financial supporter, the United States. By 1969, the militant Islamist Fatah had gained control of the Palestinian Liberation Organization. It reinforced a commitment to revenge terrorism and was instrumental in triggering the Egyptian attack on Israel in 1973. Briefly united against Israel's supporters, the Arab oil producing nations embargoed oil shipments to the United States and the Netherlands.

By the late 1960s the American, Dutch and British oil companies had begun to lose their ability to control exploration and production in the Middle East. Iraq nationalized its oil industry in 1972. Saudi Arabia purchased the assets of Aramco in steps from 1973 to 1980. Big oil also lost its control over the price of oil. World oil shortages and the declining value of the dollar[18] during the Yom Kippur crisis of 1973 gave producer nations the opportunity to increase the price of oil from and average of $3.29 per barrel in 1973 to an average price of $11.58 per barrel in 1974, an increase of 252 percent. Real and perceived shortages

---

18 History repeats. In 2004, the declining value of the dollar will again prompt OPEC to raise oil prices in order to maintain a revenue parity with 2003. Remember, oil prices are quoted in dollars. If the value of the dollar declines, the price of oil must increase in order for producer nations to maintain the purchasing power of their income.

induced by the Iranian crisis of 1979 allowed OPEC to further increase the price of oil to an average of $30.03 per barrel, an increase of 159 percent from the average price of oil in 1973.

It may seem odd, but increased Middle Eastern oil production actually made countries like Saudi Arabia more dependent on the west than ever. The United States, Great Britain, France, the Netherlands, and Germany supplied technicians, financing, equipment, arms and a ready market for the oil that funded their wealth. Saudi Arabia has thus been a strange mix of walled western compounds set apart from a closed Saudi social structure.

Iraq

By the mid-1970s, Saddam Hussein had developed a political following in Iraq. In July of 1979, he engineered the resignation of Hasan al-Bakr and assumed the role of President. A ruthless dictator, he entangled Iraq in an eight year border war with Iran in 1980 that cost the lives of over a million people. During that war he received arms and aid from the United States, Great Britain, France, Germany and Russia. Confident of his political strength, he invaded Kuwait in August of 1990. It was a bold move to end an oil field dispute with Kuwait and to expand his control over the oil reserves of the Middle East. This, of course, alarmed the Royal Family of Saudi Arabia, and they essentially agreed to finance much of the ensuing "Dessert Storm" that freed Kuwait in 1991. What followed was eleven years of fruitless wrangling over weapons of mass destruction while Saddam clandestinely sold oil to anyone who wanted it. Despite opposition by France, Germany, Russia and China, the United States and its allies invaded Iraq in March of 2003. The war against Iraq's military forces was over by mid-April.

But the war against guerrilla forces had just begun. Anti-American demonstrations erupted on April 18, 2003. Sporadic attacks on coalition forces have continued ever since. Iraqi Islamic extremists, backed by money, weapons and personnel from Iran, Saudi Arabia and other Middle Eastern countries, have made Iraq the battle ground of their war against western "Infidels". With an inexhaustible supply of suicide bombers, adequate funding from Islamic "charities", and the moral support of Muslim clerics, the Islamists fervently believe they can drive the west out of Iraq.

Oil certainly dominates Iraq's economic potential, traditionally providing up to 95 percent of the country's foreign exchange. Yet Iraq has done a very poor job of developing its oil reserves. Production was drastically curtailed by the damage inflicted during the wars with Iran and Kuwait. Prior to the war with Iran, oil production had reached 3.5 million barrels per day. Since that time, two debilitating wars, U. N. sanctions after 1991, Saddam's production decisions, capricious oil field development deals, and poor maintenance have combined to reduce Iraq's production capacity to approximately 2 million barrels per day. Sabotage and terrorism continue to constrain production.

As of February, 2004, most of Iraq's oil exports were through the port of Basra. The northern Kirkuk oil field is located in a predominantly Kurdish and Turkmen area, where political conflict overlays tribal and religious challenges to consistent oil production. The Kirkuk pipeline, which theoretically has a capacity of over a million barrels a day, needs extensive repairs in order to exceed its current (sporadic) volumes of 300,000 barrels per day. The new Gulf oil export terminal at Khor al-Amaya has been operating at less than a third of its 300,000 bpd capacity due to pumping limitations.

Information about Iraq's reserves tends to be speculative. Current folk lore pegs Iraq's reserves at 112.0 Bbl. But that number has been around since the 1980s and excludes the production of oil during the intervening years. Pessimists have concluded that Iraq has less than 40 billion barrels left. On the other hand, despite finding at least two giant fields, Saddam failed to fully develop his oil resources. They are now open to further exploration and production (assuming political stability). High end estimates exceed 200 Bbl of high-grade, cheap to produce, oil.

And so there is competition for exploration and production rights. The British sent troops into Iraq in 1918, drew the colonial boundaries of Iraq, and struggled with the United States to control the flow of oil up to 1939. After the second world war, the United States used its newly acquired political power in the region to encourage investment in the Iraq Petroleum Company (IPC) with British and French partners. The IPC coordinated its production and pricing policies with the Anglo-American oil cartel. It had a virtual monopoly on Iraq's oil, frequently restricting production in order to increase its profits elsewhere. This so irritated the Iraqi government that it nationalized the company's assets in 1972. The Americans and the British have been locked out of direct (but not indirect) participation in Iraqi oil fields ever since.

France was vocal in its opposition to the 2003 war in Iraq. The French were against the war for both political and business reasons, including a desire to reduce America's international influence, the loss of lucrative oil exploration and oil production contracts with Hussein's regime, and potential losses from uncollectable debt. More than 60 French companies provided over 22 percent of Iraq's imports. The annual trade value exceeded $1.5 billion. TotalFinaElf had negotiated a very lucrative $4 billion deal to develop the 26 billion barrel Majnoon and Bin Umar fields. The Iraqis would also have purchased vehicles, arms, communications equipment and engineering services from France.

Russia also has economic and political interests in Iraq. It sold an estimated $1.6 billion in goods and services to Iraq in 2002. Iraq was on the verge of signing a $40 billion deal for oil exploration, production, and infrastructure maintenance. The proposal included 67 new projects in southern and western Iraq, including Suba, Luhais, and Rumaila. LUKoil was pursuing a $3.7 billion, 23-year contract to rehabilitate the 15 billion-barrel West Qurna field in southern Iraq. Additional oil field contracts, worth over $100 million, were in process. Like France, Russia is apprehensive about its future role in the Middle East, wary of American intentions and concerned about Iraq's future as a trading partner.

A partnership between China National Oil Company and China North Industries Corp. had negotiated a multiyear deal for oil exploration in the Al Ahdab oil field of southern Iraq. Chinese companies sold arms, electronics and communications equipment to Iraq.

From 1981 to 2001, according to the Stockholm International Peace Research Institute (SIPRI), Russia supplied Iraq with 50 percent of its arms, China sold Iraq 18 percent of its arms, France was responsible for over 13 percent of Iraq's arms imports, and Germany sold arms and "dual use" technology to Iraq.

European banks, including BNP Paribas and Crédit Agricole of France; Deutsche Bank and Hypovereinsbank of Germany, Banco Bilbao of Spain, Rabobank of the Netherlands and Credit Suisse Group of Switzerland, were competing against each other to handle the money Iraq made from its oil sales under the UN food for oil program[19.]

But this commerce and banking relationship cost France at least $5.5 billion in unpaid debt. Saddam owed Russia $8 billion and Germany another $5 billion. CNN reported that Iraqi owes $40 billion to 19 creditor nations. Total debt exceeds $65 billion[20].

Rebuilding Iraq will take time and money. War and neglect have decimated schools, hospitals, roads, bridges, and commercial buildings. Saddam Hussein failed to maintain the infrastructure of Iraq's oil industry. Assuming a relatively stable political environment, the effort will take 5 to 10 years and $200 to $350 billion USD. If Iraq is able to ramp up its oil production to 3.0 million barrels a day, and the market supports an average 10 year price of $38 per barrel, then Iraq will be able to generate oil sales revenues of $416 billion from 2004 through 2013. Given the cost of exploration, production, processing, and

---

19 European Banks Jostle for Iraq's UN Contract, By Carola Hoyos, reference Financial Times, February 19, 2002, Global Policy Forum, http://www.globalpolicy.org

20 The information for this discussion about Iraq came from multiple sources. An excellent resource of additional information may be found at The Heritage Foundation, http://www.heritage.org/research/MiddleEast/issues2004.cfm. Also check out the sources listed in the References section of this report.

transportation, Iraq will need to sell a lot more oil to pay off its reconstruction debt.

The Europeans, aside from their commercial interests in Iraq, understood the consequences of America's invasion in 2003. The war would serve to unite the otherwise independent and frequently feuding Islamic organizations against the west. Shiite and Sunni would come together in their common belief that the infidel must be driven from Iraq. Ignoring this possibility was a major strategic blunder for the United States.

German, French and Russian opposition to America's invasion of Iraq was also motivated by an internal political reality. Muslims constitute a growing proportion of the European population. Low birthrates among traditional European ethnic groups are being partially offset by high birth rates among Muslim inhabitants. While this shift of bedroom virility is taking place, Western Europe is also experiencing an extraordinary influx of Muslims from Turkey, North Africa and the Middle East.

Europe has 727 million inhabitants. Of these, 269 million are Catholics, 107 million are Protestants and 53 million are Muslims. There are 5.7 million Muslims in France, 3.1 million in Germany, and 2 million in the U.K. Some 25 million live in Russia and the southeastern part of the FSU. European Muslims, particularly those located within the European Union, are becoming a more powerful political force. Although they come from different countries and religious preferences, they all share a sympathy for Palestine and Palestinians. And unlike most of their Arab brethren, growing numbers of Europe's Muslims can vote.

Tension and controversy are increasing. The newly enfranchised Muslims in Europe are vocal, virile and vigorous.

Motivation

My consulting and marketing career took me to over 20 countries. One of the tricks to foreign travel is to arrive a day or two early in order to acclimate to the new time zone. It also affords an opportunity to take a stroll through the city, visit the local attractions, and enjoy a few moments of conversation (however awkward) with the local residents.

From these experiences I have learned that children are much the same in any country. Inquisitive, enthusiastic, shy or outgoing, always in motion, given to laughter and squeals of delight as well as moments of tears, always inventing games to play - they are a delightful reminder of human potential.

Most of the parents I met also shared many characteristics. They were proud of their children, attentive to their families, eager to improve the comforts of daily life, and anxious to avoid divisive confrontation. They typically were faithful to some higher belief system

- religious conviction. They wanted a stable cultural environment so they could get on with the chores of daily life without fear or disruption. If all the peoples of the world espoused these values, there would be no war because the ravages of conflict interrupt the harmony of life.

Unfortunately, poverty exists. It breeds frustration and anger. Great wealth exists. It breeds condescension and overindulgence by those in power. Although most of the people within any country may yearn for the harmony of a peaceful life, the economic and cultural environment within which they exist may preclude its realization.

Such is the case with most of the Arab universe. Ostentatious wealth exists beside abject poverty. For the frustrated, religion is an opiate that dulls the pain of want and provides a channel for anger. This is a culture of stark contrasts. Most Muslims have a preference for the doctrines of compassion that can be found in the Koran. Family, clan and religion are important elements of human psychology. But some nations (by Western standards) cling to antiquated cultural customs[21.] Masculine conservative thinking dominates every aspect of daily life. Women are subservient to oppressive restriction. Religious education and cleric exhortation fuels a continuing stream of anger against western citizens and institutions. Among Islamist zealots, deep seated hatred and distrust have exploded with insidious energy. In their view, the enemy is clearly identified, badly disorganized and endemically corrupt.

Now then. Do we believe these cultural circumstances will actually impact world oil production?

YES.

Oil brought the Jewish/Christian/secular West into Arab countries on an intimate basis. Arabs now find themselves living with emissaries of a foreign civilization whose presence they distrust[22]. Radical Islamists and the teeming masses of poverty stricken Muslims perceive little benefit from the production of oil. Westerners are frequently branded as "infidels". Western cultural influence is being challenged with violence.

The Wahabbis of Saudi Arabia have rejected all liberal forms of Islam. They espouse a strict and repressive interpretation of Islamic law. They believe that anything less than a doctrinaire adherence to the law will result in damnation. The only sure way to get to heaven is through martyrdom. According to the 13th century philosophy of Ibn

---

21 For additional information, read Jan Goodwin's book, Price of Honor, Muslim Women Lift the Veil of Silence on the Islamic World, Little Brown and Company, 1994; or Ibn Warraq's book, Why I Am Not A Muslim, Prometheus Books, 1995.

22 Recommended reading: A History of the Arab Peoples, Albert Hourani, The Belknap Press of Harvard University Press, 1991

Taymiyah, it is the obligation of a good Muslim to die for the cause of Allah - "to die a Martyr for the unification of all the people, in the cause of God and his word, is the happiest, best, easiest and most virtuous of deaths." There can be no compromise. Muslim leaders who seek resolve political conflict through peaceful negotiation with the enemies of Allah commit apostasy - the abandonment of one's faith in Allah and the laws of the Qur'an. In the eyes of the extremist, apostasy is punishable by death.

With Saudi financial support, the Wahhabis are determined to spread their brand of Islam throughout the world. There is a radical element that propels us toward confrontation. The extremists have adequate financing. They operate an efficient and effective propaganda campaign. Their activity has spread throughout the world. Passionate and deeply held religious beliefs along with centuries-old grudges provide the emotional justification for brutal revenge.

In a corrupt world, Islam is the solution.

Jihad

Prophets play a primary role in the practice of Muslim beliefs. They usually take a long term view of history, provide a voice for Islamic beliefs, and have a lasting influence over Islamic thought. In his own mind, Osama bin Laden is (or was) such a prophet. He has a long term view of Muslim influence and a vision of how Islam fits within the scope of world history. Osama bin Laden believes that it is time for Islam to expand across the earth, cleansing humanity of its infidel decadence. For him, and his followers, their losses in Afghanistan and Iraq are merely setbacks in a world of permanent war against Christian and Jewish evil[23].

Osama believes that America's intervention in the Middle East is nothing more than an attempt to impose Jewish control over the Palestinians. The Jews need the support of their Christian brothers in order to dominate the Arab Peninsula. Muslim terrorism is praiseworthy and fully justified because it is directed at the enemies of Allah. In this war, there is no reason to differentiate between civilian and military targets because civilians support evil. The corruption of Western influence must be driven out of every Islamic nation. All nations must embrace Islam in order to achieve spiritual and fraternal harmony. Although Osama bin Laden is not the administrative leader of Muslim terrorist activity, he is certainly a heroic spiritual icon for

---

23 A very revealing interview with Osama bin Laden conducted in May, 1998 can be found at the PBS.org WEB site. It was held at bin Laden's remote mountaintop camp in southern Afghanistan two months before the truck bombings of the U.S. embassies in East Africa.

militant Muslims. If there were an election in Saudi Arabia, he would win.

Osama is not alone. Islamic fundamentalists have a deep and fervent belief they are acting in the service of Allah and the prophet of Islam. The purity of Islam must not be corrupted by Western cultural values. Since neither secular government nor socialism has been able to improve the economic and social status of Muslim nations, the only plausible solution is to return to theocratic political systems. Islamic fundamentalists reject the ideals of Western democracy. In their vision of a purified world, nations are ruled by a doctrinaire theocracy. Only clerics have the proper training to rule. There is no religious freedom. There can be no freedom of speech. There can be no challenge to the absolute truth. Human freedom must be restricted to those actions and ideas that are permissible under a strict interpretation of Islamic law.

> There is only one path to peace. Islamic victory.

Despite its losses, the extensive al Qaeda organization retains its military, security, economic and media capabilities. It has shown a remarkable ability to incite anti-American propaganda. The men who run this spectral organization are well-educated and very intelligent. They understand the power of public relations and political alliances. They have infiltrated or control multiple Islamic non-governmental associations, and have successfully infiltrated several of the mainstream national political parties.

Europe remains al Qaeda's forward position for logistics, financing, and recruits in its war against the West. It laid the groundwork for its infiltration into Europe in the early 1990s. Most of the early militants were members of a fiery puritanical Salafist sect of Islam located in North Africa. The Islamic Salafists want to restore the true essence of an Islamic society, including establishing a caliphate and the imposition of Islamic law. With the help of other Islamic militant groups, the Salafists are perfectly capable of destabilizing the oil-producing regions of northwestern Africa.

The point, of course, is that when we consider what is happening in Iraq, Afghanistan, Pakistan, Syria, Iran, Saudi Arabia, Indonesia, and elsewhere, the war against terrorism is far from over. We have merely entered another phase of a movement that has a long history. There are active al-Qaeda cells or camps in Yemen, Jordan, India, Egypt, Qatar, Syria, Saudi Arabia, Morocco, Pakistan, Palestine, Sudan, Chechnya, Kashmir, Iraq, the United Arab Emirates, Spain, France, Britain, Germany, Italy, the Netherlands, Indonesia, Bosnia, Australia and - of course - the United States. Any nation with a sizeable Muslim population has become a target for infiltration. That's probably 60 countries. Recruits are not hard to find among the disillusioned and the alienated. Radical solutions to complex problems appeal to the young. Poverty and ignorance feed human hatred. Extremist interpretations of Islamic theology provide the excuse for vengeance.

Robert Baer, in his book "See No Evil"[24] describes a walk he took through the Muslim book stalls of London. Tract after tract advocating Islamic violence against western culture. A deep, uncompromising hatred of the United States. The assumption that an Islamic Jihad (holy war) against the west is already underway. Multiple groups of terrorists venting their anger in the name of Allah. This is the reality in Iraq, Iran, Kuwait, Saudi Arabia and the United Arab Emirates - where most of the Middle Eastern oil reserves are located. It is the mantra of extremist Muslims living in Syria, Pakistan, Afghanistan, Azerbaijan, Kazakhstan and the Eastern oil-producing areas of the Russian Federation.

The Islamists have a long term view of history. Jihad has been continuing for 1400 years. It has been revived by the Mujahideen who are now gaining ground against the forces of evil. The conflict between Muslims and Christians will continue until Islam has triumphed all over the world.[25]

For the purposes of this report, it is important to recognize the potential impact of Islamic fundamentalism.

> Muslim cultural beliefs influence government policy
> in nations that own 798 Bbl - or 76 percent -
> of the 1,047.7 Bbl of Proven Reserves
> tabulated by BP at the end of 2002.

At this point in history, it is less important that these reserves exist. It is far more important to understand if they will be available for production.

Islam is the solution. Jihad is the way.

Oil is a weapon of war.

---

24 See No Evil: The True Story of a Ground Soldier in the CIA's War on Terrorism, Robert Baer, Crown Publishers
25 Go to your favorite search engine on the Internet. Type in jihad and Christian. Read a sample of the commentary on the contemporary confrontation. Yes. This is a religious war. Just ask the Jihadists.

Collision Course

Where do organizations like al Qaeda and the Muslim Brotherhood get their money? From multiple organizations, individuals and several governments. But the primary financial resource is Saudi Arabia - the keystone of world oil supplies.

There is ample evidence, for example, that organizations such as the sophisticated and well funded Muslim World League, a semi-official agency of the Saudi government, has been financing the spread of Wahabbi Islamist doctrine[26.] Imams in Saudi Arabia's larger Mosques - which operate with the approval of the Saudi government and are funded either by the government or the Royal family, routinely call for the defeat of Jewish and American interests in the Middle East[27]. Saudi Islamists have apparently joined with radicals from Iran[28] to disrupt the democracy effort in Iraq. Although an examination of the link between Saudi charity and Islamist Jihad is well beyond the scope of this report, this activity does raise a serious question:

What is their motivation?

If the al Sa'ud are paying protection money to dissident Arab groups in order to protect their political power, then they face an uncertain future. Eventually they will be overthrown and in the process, the violence of the Political Crisis scenario described in this report is not merely possible - it becomes highly probable. If, on the other hand, the Al Sa'ud family is dedicated to funding the spread of Islam, then the Wahabbi/Royal Family relationship, however tenuous, is likely to stay in place. Saudi oil production is more likely to gradually decline as western technicians are purged from Saudi soil and Saudi Arabia reduces production as described in the Production Crisis scenario.

The Jihadists admit that perpetrating a civil war in Saudi Arabia could backfire. On the one hand, Saudi Arabia is a holy land. The center of Muslim geography is in Mecca. The strongest support for Jihad and Islamic fundamentalism exists here. Saudi Arabia is a resource for manpower, funding and bases. If the Royal family is threatened, it may seek to retain its power by asking for American military intervention. The resulting conflict would threaten the security of al Qaeda's

---

26 Two related Saudi agencies, the World Assembly of Muslim Youth and the International Islamic Relief Organization, have also been implicated. Ample references to their activity can be found on the Internet.

27 Read the complete text of "Kingdom of Venom", by Joel Mowbray, From the WEB site of the National Review Online, May 29, 2003, www.nationalreview.com.

28 Read Eleana Gordon's article, "Democratizing Iraq, The U.S. needs to do much more to make it happen", in the National Review Online, December 01, 2003, www.nationalreview.com

operations. On the other hand, there are many Islamists who believe that Jihad against the west must be fought on Saudi soil where they have the numbers to defeat the armies of the west. A victory would prevent further American intervention in the Middle East and weaken western resolve to bring democracy to Iraq.

The decision underscores the depth of Wahhabi Islamist resolve to remove western influence from the Islamic world. Those who favor aggression are in control of Jihad. Holy war demands bloodshed and sacrifice. Compromise only signifies weakness. Frustration, anger and hatred dominate current beliefs.

There is also this mentality in Saudi Arabia that oil has been a curse. For many Saudis, including some members of the royal family, taking the kingdom's oil off the world market - even at the risk of destroying their own economy--is an acceptable alternative to the status quo. Sure. Most Saudis could not survive without oil money. But they believe they can.

And that is all that matters.

By restricting the world's supply of oil the Islamists accomplish three things:

1. it will devastate the economies of the western nations;
2. the price of oil will increase, giving them more money to fund Jihad; and
3. they will be able to conserve their oil as a resource for funding the future of their regime.

The clouds of revolution are gathering. Will Saudi Arabia self destruct? Or will the Wahhabis channel their vengeance into attacks on other nations? Either way, Saudi Arabia is on a collision course with the industrialized nations of the world.

For oil production, it will not be business as usual.
The Saudi oil infrastructure is far too vulnerable to political chaos.

Oil shortages are inevitable.

Islamist Reality

It does not matter if we favor or oppose the war in Iraq. It does not matter if we favor Jew or Palestinian, Muslim or Christian, Democracy of Theocracy. It does not matter if we are of liberal or conservative ideology. For the purposes of this report it is enough to understand one simple fact:

Islamist Jihad exists.

It would be a fatal mistake to underestimate its global reach. Islamist political power is certain to disrupt oil exploration, production and transportation in the Middle East, Northern Africa and the Muslim states of the former Soviet Union (FSU).

Continuing and predictable oil production is vital to sustaining the stability of the world's petroleum markets and thus the health of the world's economy. Peace in the Middle East must therefore be the focal point of an international effort. If consumer nations are going to be able to develop fair and equitable agreements with producer nations, then both sides will need to represent stable political structures.

Is there sufficient political stability in the Middle East to permit accommodation and compromise? Can we build a bridge between wary cultures? Can we encourage reform as a byproduct of human interaction? Will we have the patience?

If the international community fails to deal with the existing conflict, then how will this vigorous display of Islamic rebellion and virility play out? It's all downhill. Worldwide. Confrontation in the Middle East, North Africa, and the Former Soviet Union will exacerbate the problems of oil depletion.

Jihad is reality. An oil crisis will plunge our planet into eternal economic decline.

Chapter 6 HIGH PROBABILITY

The following scenarios challenge cherished beliefs. Skeptics will go out of their way to deny their possibility. We pray they are right.

Introduction

The No Change and Best Case scenarios both make the broad assumption that there are no cultural or political impediments to oil exploration and production. But is this realistic? The Middle East has a long history of conflict - Jew versus Muslim, Muslim versus Muslim, Muslim versus Christian, Muslim versus Infidel, and nation versus nation. Furthermore, the cultural challenges described in the preceding chapter are not going away. As we are see on television every day, they are real, present, and will continue for the foreseeable future. It is most unfortunate, but we humans do not have a credible plan to bring peace to the Middle East and we remain divided about the issues of cultural change that must take place before the Islamist threat can be defused.

The following Production Crisis and Political Crisis scenarios attempt to evaluate the impact of the Islamist threat in economic terms. We essentially start with the oil production and consumption data presented in the Best Case scenario, make minor adjustments to reflect real world labor and political conditions, and then factor in the economic impact of this very real cultural conflict.

The Production Crisis Scenario

The Production Crisis Case study makes two very reasonable assumptions.

1. First, it assumes that the existing patterns of Islamist influence persist throughout the Middle East. There is little change to the decades old violence between Jew and Palestinian. Iraq remains politically unstable and occasionally violent. Terrorism continues in Western Europe, the United States and elsewhere. Yet despite these problems, the political landscape of the Middle East remains relatively unchanged.

2. Second, this scenario assumes that a combination of Islamist influence and resource depletion stalls OAPEC production capacity at a maximum of 25.4 million barrels of oil per day. Most of the additional Saudi production is based on maximizing the output of existing production facilities. By 2008, these have been converted to full time production. Iraq contributes up to 3 Mbl per day by 2010. Production in the other OPEC nations follows existing depletion patterns. Total OAPEC output increases from 20,973 Mbl per day in 2002 to 25,402 Mbl in 2013 - the year of peak OAPEC production.

Islamist influence limits the use of foreign technicians and engineers to sustain production. Exploration is lethargic. Depletion gradually restricts production. The existing infrastructure in key producer nations such as Saudi Arabia and Iraq gradually deteriorates because of poor maintenance. Output is erratic and demand exceeds supply. By 2011, OAPEC production will no longer cover shortages that occur in other oil producing regions.

The results: **Relative calm in the Middle East can not prevent a Production Crisis**.

In this scenario we also assume that total North American, EurAsian and ROW oil production is reduced by technical factors and labor conflict to a volume that is approximately 2 percent less than these regions produce in the Best Case Scenario. (A decrease of only 2 percent is probably optimistic). World oil production is lethargic and disorganized (highly probable).

As a result of these assumptions, world oil production peaks at 32.9 Bbl in 2021. North American oil production falls by 36 percent over the Forecast Period (highly probable). Middle Eastern production increases by 8 percent (possible). EurAsian and ROW production increase by 25 and 40 percent, respectively (probable). However, even with these relatively optimistic assumptions, total world oil production, at 616.8 Bbl over this 20 year period, would be *6 percent less* than the production assumptions of the *Best Case* Scenario. As shown in Figure 10, production is essentially flat from 2012 through 2022 at 31.7 to 32.9 Bbl per year.[29]

In our Production Crisis scenario, a series of oil shortages will begin in 2009 because short term demand (which tends to be volatile) will often exceed the oil industry's ability to increase (or even sustain) annual production.

---

29 For purposes of comparison and information, Figure 10 includes actual historical data from 1970 - 2002.

Figure 10
World Oil Production
Production Crisis

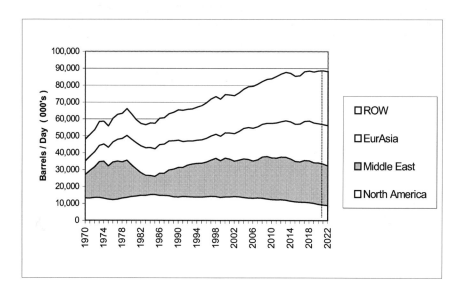

World oil production has increased 17.5 percent from 28.0 Bbl in 2003 to its peak in 2021. Yet despite the net increase in world oil production, consumer region economic activity is sharply curtailed by the lethargic growth in oil production experienced after 2012. Competition for oil resources propels conflict. By 2016, all nations begin to hoard their reserves for future use. Although the production crisis begins to be felt in 2014, it really hits in 2019 when the Rate of Change for world consumption turns negative (Figure 11). Compare the projected Rate of Change for 2003 - 2022 with the actual Rate of Change we experienced from 1970 - 2002. Our projected volatility is not all that unusual. Indeed , it could be much worse, driving us deeper into recession than the assumptions used as the basis for this scenario would predict.

> The world oil crisis intensifies from 2014 through 2019,
> culminating in a permanent economic recession
> as national wealth evaporates.

North American oil consumption only grows at a Compound Annual Growth Rate (CAGR) of 1.0 percent through the Forecast Period (versus

1.3 percent in the Best Case Scenario). Western Europe struggles with a zero percent growth in consumption (versus a .7 percent annual growth in the Best Case Scenario). Asia/Pacific regional consumption drops to a CAGR of 1.7 percent (versus 2.8 percent in the Best Case scenario). ROW consumption drops by a Compound Annual Rate of Decline (CARD) of minus .2 percent (versus an assumed increase of .3 percent per year in the Best Case scenario). Sure, we can change the economic calculations about which nations get the oil, but all regions will suffer. If the world's natural oil demand has a CAGR of 1.8 percent, and a Production Crisis reduces world consumption from 2003 through 2022 to a CAGR to .8 percent, every nation will take a hit.

Figure 11
Rate of Change
World Oil Consumption
Production Crisis

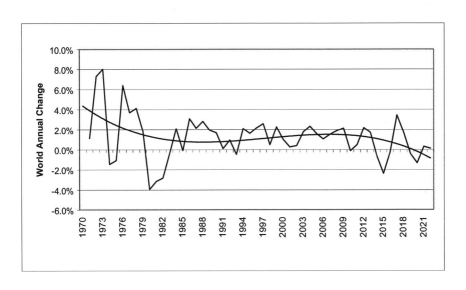

All of our scenarios include the assumption that there will be an increase in the world's natural demand for oil of 1.8 percent per year from 2003 through 2022. In Figure 12, we can see that annual oil production falls below the natural rate of demand for most of the Forecast Period. Sporadic oil shortages occur. When oil production does exceed current consumption, natural demand quickly absorbs the excess volume of oil. Unsatisfied demand combines with escalating exploration and production costs to drive up the price of oil to $75.00 per barrel by 2022. Compare this trend with the very large price increases that occurred during the oil shortages of the 1970s. Our price projections are probably conservative.

Figure 12
Oil: World Natural Demand, Production and Price Production Crisis

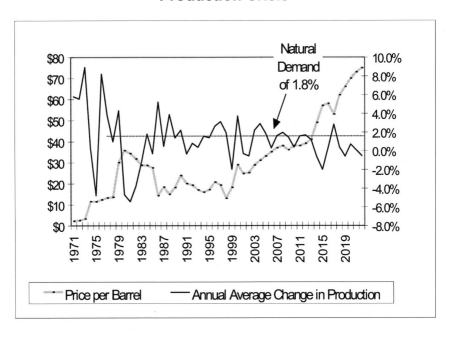

North American demand exceeds consumption in 10 out of 20 years from 2003 through 2022. Western European demand exceeds consumption in 11 out of 20 years. For the Asia/Pacific region, it is 17 out of 20 years, and for the ROW nations, it is 15 out of 20 years. Shortages, combined with higher prices for a barrel of oil, will drive up the rate of inflation. By 2020, North American inflation will exceed 6 percent. Based on a very conservative calculation of gasoline prices, a gallon of gasoline in the United States will cost at least $4.03 by 2022 - if you can find it. Gasoline prices in Europe and the Asia/Pacific Region will be substantially higher. In France, for example, it will exceed $2.30 US per liter.

In Figure 13, we can see that rising rates of unsatisfied demand are instrumental in driving up the price of oil. We need to remember that as the price of the products made from oil (including gasoline and heating oil) increase, there will be a <u>decrease</u> in consumer demand. Hence, unsatisfied demand must be adjusted (decreased) as prices rise. In this Scenario, the persistent increase in unmet demand signals an equally uncompromising and unavoidable decrease in consumption as consumers adjust their life styles.

In our scenario, oil production has reached 32.7 Bbl per year by 2022. If there had been enough oil to satisfy consumer demand from 2003 to 2022, then world demand <u>and</u> consumption (ignoring the impact of price changes) would have grown to 39.5 Bbl of oil by 2022. But with each passing year, net average demand increasingly exceeds consumption. By 2022, annual shortages will have increased to 6.8 Bbl, or 17.2 percent of natural consumption.

However, increasingly higher prices have decreased consumer demand. They buy smaller cars and drive less. Manufacturers reduce demand by switching to alternative petroleum feedstocks. Coal and coal by-products are substituted for oil and natural gas in many energy applications. As a result of these adjustments in consumption, by 2022 world unsatisfied demand has actually declined to 12.7 percent of natural demand or 5.0 Bbl of oil. Never-the-less, that shortage is large enough to drive a continuing and relentless escalation of world oil prices.

Figure 13
World Oil Market: Adjusted Unsatisfied Demand Vs. Price
Production Crisis

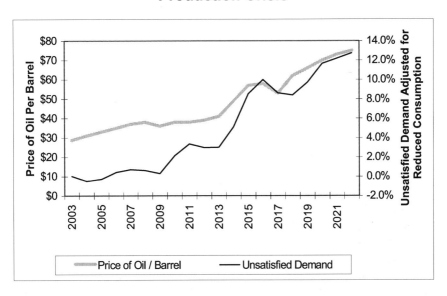

Given the assumptions of the Production Crisis, and after adjusting demand for oil shortages and higher prices, we can expect that during the Forecast Period the world will experience a cumulative shortage of over 700 billion gallons of gasoline.

North American oil consumption peaks in 2018 at 28.9 Mbl per day (Figure 14). Western European consumption peaks in 2013 at 14.5 Mbl per day. The Asia/Pacific region peaks at less than 31 Mbl per day, far less than the 38 Mbl per day we calculated for the Best Case Scenario.

Figure 14
North American Oil Consumption
Production Crisis

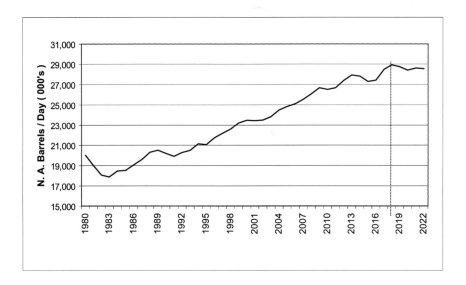

As one would expect, competition for available oil resources drives up the rate of inflation. Within the United States, world oil prices - along with the inflationary impact of declining productivity - drive up the annual rate of inflation to over 6 percent by the end of our Forecast Period (Figure 15). The average rate of Inflation will increase from the historical rate of 3.14 percent experienced during the 1983 - 2002 timeframe to 4.52 percent during our 2003 - 2022 Forecast Period. The rate of inflation would be higher if it were not for the impact of debt defaults, the deterioration of the financial markets, and higher unemployment - all of which are deflationary.

Figure 15
World Price of Oil Vs. U. S. A. Rate of Inflation
Production Crisis

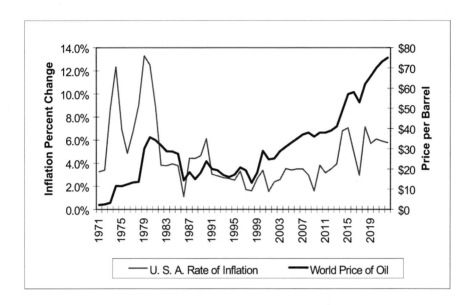

Figure 16 and 17 illustrate the impact of a Production Crisis on oil consumption, Gross Domestic Product (GDP), and Unemployment. As shown by the data used for the Best Case Scenario, U. S. GDP is declining. If we experience a Production Crisis that mimics the assumptions used in our model, then GDP will actually dip into negative territory by the end of our Forecast Period. Overall, a Production Crisis would reduce average annual U. S. GDP growth from 6.01 percent in the 1983 - 2002 timeframe to 3.07 percent during the 2003 - 2022 Forecast Period, a decline of almost 50 percent. If this trend continues, the economy of the United States will stop growing before 2030.

The impact on unemployment will be tough to manage. By 2015, there will be significant layoffs in any industry that is directly tied to the consumption of oil (auto, distribution, trucking, etc.) and peripheral layoffs in those industries that are tied to economic growth (retail, housing, etc.). Although the average annual rate of unemployment for the Forecast Period (6.47%) is only slightly higher than the 1983 - 2002 historical rate of 6.02 percent, there will be a sharp trend upward in the later years of the Forecast Period. By 2022, unemployment will exceed 8 percent.

Figure 16
Rate of Change: U. S. A. Unemployment and GDP
Oil Production Crisis

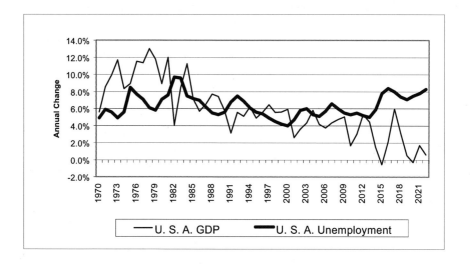

Figure 17
Rate of Change
N. A. Oil Consumption and U. S. A. GDP
Oil Production Crisis

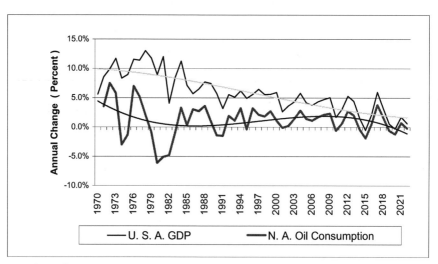

If you live in Western Europe, China, India or Japan - take another look at the Charts presented for the Production Crisis Scenario. If a

Production Crisis brings oil consumption growth down from a CAGR of .9 percent (1983 - 2002) to zero percent (2003 - 2022) in Western Europe, then *the European Union will have to struggle with economic and social forces that could easily tear it apart. Economic and social circumstances will be far worse in China, India and Japan.* Emerging expectations for increased well being will be cut short if the growth of consumption declines from a regional average of 4.0 percent (1983 - 2002) to 1.7 percent (2003 - 2022). The resulting GDP, Unemployment and Inflation trends of these countries will be similar to the ones identified for the United States.

If we experience a Production Crisis,
all nations will suffer economic deprivation
and growing social tension.

Conclusion: If the industrialized nations of the world fail to work together, if we fail to deal with the cultural challenges of the Arab world, if we fail to find a way to constructively manage a Production Crisis, or if we fail to pursue the Recommendations found in Chapter 9 of this report, then this scenario - or one that has similar characteristics - has a 70 percent probability of becoming our future.

My fear, indeed the fear we all should share, is that political squabbling and imperious attitudes will take precedence over common sense. Stubborn ideology will prevent constructive action.

If we fail to work together to solve our planet's energy problems,
then international conflict is inevitable as a Production Crisis
tightens its noose around the windpipe of economic growth.

Perhaps the economic characterization of the Production Crisis scenario is much too optimistic.

The Political Crisis Scenario

The first three scenarios assume a relatively stable political environment in the oil producing regions. Although the inevitable conflict between Palestinian and Jew, Shiite and Sunni, Curd and Iraqi, Wahabbi and al Sa'ud, Coalition and radical, will continue, there is no upheaval that changes the character of any government - Middle Eastern, African, Asian, or elsewhere. Oil production problems are primarily caused by he realities of oil field geology, geography, engineering, exploration, production, and logistics. In the Production Crisis Scenario, a combination of Islamist influence and lethargic oil field development merely reduce oil exploration and production in the Middle East.

But is this realistic? What happens if extremist activity disrupts the political environment?

You can put me in the camp that believes America's relationship with the House of Sa'ud borders on the absurd. Presidents going back to Franklin Roosevelt have sought to develop and maintain a close relationship with Saudi Arabia by promoting an agenda of insider deals and expensive gifts. It's the personification of the "old boys club" approach to diplomacy with favored contracts tied to alleged corruption and payoffs. According to Robert Baer, the CIA should have understood the Wahhabi terrorist threat. But the agency was told to "back off" by the State Department. Investigating the House of Sa'ud was not politically expedient. The identification of "Arabs" as a terrorist threat was not "politically correct".

It's nice to be an idealistic social liberal. Our relationship with the House of Sa'ud has been naughty. This stuff should not be happening.

But it's also important to understand how the world actually works. Because of this relationship, Saudi Arabia has come to America's rescue on multiple occasions to make sure we have enough oil to sustain our lifestyle and economy. They used their excess production capacity (and made a lot of money) when they broke the OPEC oil embargo of 1974. They again used their excess capacity to stabilize the international oil market during the protracted Iran - Iraq war of the 1980s, the Gulf War of 1990 - 1991, and after the shock of 9/11/2001. Saudi Arabia has the world's only really useful surplus production capacity -- two million barrels a day. This keeps the world market liquid. And it gives them a very loud voice in oil price discussions. Because oil is an international commodity, even countries that don't buy Saudi oil would be vulnerable to economic chaos if the flow of oil were disrupted.

Unfortunately, the House of Sa'ud and Saudi oil production is frighteningly vulnerable to attack. Saudi Arabia shows signs of coming apart at the seams. That puts its more than eighty oil and natural-gas fields, and more than one thousand working wells in jeopardy. Eight oil fields, including Ghawar - the world's largest oil field on land- and As

90

Saffaniyah - the world's largest offshore oil field - account for almost half of Saudi Arabia's oil reserves. The world's largest oil-processing facility is at Abqaiq. All petroleum originating in the south is pumped to Abqaiq for processing. Almost seven million barrels a day. The Abqaiq complex, about twenty-four miles inland from the northern end of the Gulf of Bahrain, is vulnerable[30]. So are the oil tanker loading terminals at Ras Tanura and Ju'aymah. And Pump Station No. 1, the station closest to Abqaiq, which sends oil uphill, into the Aramah Mountains, so that it can begin its long journey across the peninsula to the Red Sea port of Yanbu. And of course the Qatif Junction manifold complex, which directs the flow of oil for all of eastern Saudi Arabia.

Take out a half dozen facilities - key points in the chain of oil processing facilities and pumping stations, and the world's oil production capacity would drop by 6 to 8 Mbl per day. That may only be 8 to 10 percent of daily production. It doesn't sound like much. But my work with the oil depletion model confirms that it takes only a small change - less than 3 percent of daily production - to impose a very large hardship on the world's economy because the impact ripples through all of the consuming nations.

For months after an attack, poisonous fumes and residue along with a lack of spare parts would paralyze production. Perhaps 30 percent of the lost capacity could be restored in 5 months. Eight months of effort would bring production to 50 percent of capacity. After a year, 80 percent of lost Saudi production would be restored. Full production might take up to 24 months - IF there aren't any more terrorist disruptions. And in the meantime, the world's oil supply would be volatile. There would be no elasticity in the chain of production to buffer consumer demand. Prices would skyrocket. Economic hardship and escalating unemployment would generate a wave of political instability.

And what happens if this is a coordinated attack? One that takes out two or three other Middle Eastern oil complexes? And what happens if there is a wave of al Qaeda sympathy strikes in other Muslim countries?

Things could get very dicey.

Think it can not happen?

---

30 Information from The fall of the House of Saud, Robert Baer, The Atlantic Monthly, May 2003; and Sleeping With The Devil, How Washington Sold Our Soul for Saudi Crude, Robert Baer, Crown Publishers.

Osama bin Laden is a hero - dead or alive. During the summer of 2002, almost 80 percent of the hits on the al Qaeda Website came from Saudi Arabia. This country is at war with itself. It is torn between the mores of twelfth century philosophy and the reality of a 21st century global community. The Al Sa'ud funded terrorist groups and radical Mosques. Protection money? These same groups believe that the House of Sa'ud has failed to protect Islam from evil in Afghanistan, Palestine and Iraq. Oil money has corrupted the ruling family.

The Al Sa'ud must be punished.

The assumptions of the Political Crisis scenario attempt to quantify the impact of surging Islamic extremist activity. Religious fervor causes a political crisis in the Middle East and elsewhere. Al Qaeda or some other fundamentalist group takes over Saudi Arabian production in 2006 and begins to limit output. By the middle of 2007, sharp reductions in oil production cause a worldwide political crisis[31]. Sympathy strikes occur in all oil producing regions that have a large Muslim population. Terrorist activity disrupts the exploration, production and transportation of oil. Western military intervention has limited success in restoring production. Islamist controlled governments withhold oil production in order to punish the West.

The Political Crisis scenario makes two optimistic assumptions:

1. the confrontation does not erupt into war in the Middle East, and

2. production is quickly restored when Muslim leaders recognize the need for cash to sustain their political power.

By that time, however, aggregate international oil production will no longer be sufficient to rescue the World from a devastating economic depression.

The results: **This could also become our future**.

The impact will be felt by every nation. By mid-2008, there has been a 12 percent reduction in oil production. In the United States and Canada, environmentalist objections are swept aside in order open up additional areas for exploration and production. Although exploration begins immediately off the coast of California, the Atlantic continental shelf, and the Arctic, new production does not begin to come on-line until 2011. In this scenario, these moves will increase North American production to 89.9 Bbl of oil during the Forecast period. The net impact of this additional capacity, however, only serves to cushion the decline of North American production.

Middle East oil production takes a massive hit. Although the absolute decline is actually less than the one experienced 1979 - 1985, (Figure 18) there is insufficient spare capacity elsewhere to make up for

---

31 The dates used for this scenario are a calculated assumption for the purposes of constructing the underlying hypothesis.

the shortfall. In our assumptions, we have taken an optimistic view of Middle Eastern oil production in this Political Crisis scenario. Despite continuing conflict and the intransigence of the new Saudi Arabian regime, production recovers to 21.2 Mbl per day by 2022.

Although sympathy strikes disrupt oil production in the African and EurAsian oil fields from 2007 to 2009, further decreasing the supply of available oil, they quickly run their course. Production rapidly recovers. Sharply higher prices stimulate frenetic activity in the oil fields of the EurAsian and ROW regions. There is a brief surge in production. In the aggregate, however, world oil production over the 20 year period from 2003 through 2022 drops to a CAGR of .7 percent. This is far less than the 1.8 percent CAGR that the world needs in order to sustain its current level of economic growth.

Figure 18
World Oil Production
Political Crisis

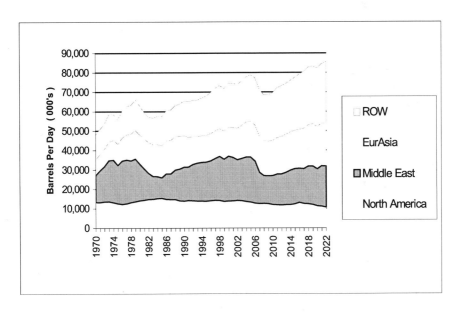

Although oil production during the 20 year Forecast Period is actually more than 85 percent of the production characterized for the Best Case scenario, and we assume that it actually recovers to 31.7 Bbl per day by 2022, this disparity is sufficient to cause a worldwide depression. If we humans are forced to endure the impact of a Political Crisis scenario, then North American and Asia/Pacific consumption would retreat for eight years. Western Europe and ROW nations -

which have little oil of their own - would feel the pain for 17 years. From 1983 to 2002 (Figure 19), oil consumption increased at an average annual rate of 1.34 percent. In our Political Crisis Scenario, it declines to .71 percent during the 2003 - 2022 Forecast Period.

Figure 19
Rate of Change
World Oil Consumption
Political Crisis

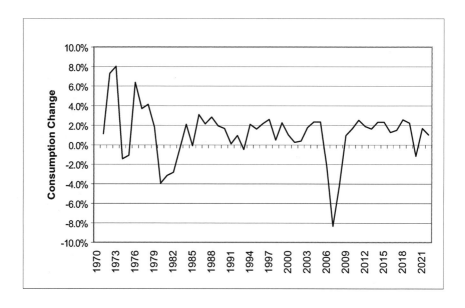

World oil prices skyrocket (Figure 20), reaching $79.00 per barrel in 2008. In the U. S. A., the average price for a gallon of gas reaches $3.90. Over the 20 year Forecast Period, total world gasoline shortages - adjusted for reduced consumption - exceed 1.2 Trillion Gallons. In the irony of numbers that attest to this crisis, world consumption does not peak during the Forecast Period because in a Political Crisis, less oil is available for consumption from 2003 - 2022. Hence, more oil is left to produce from 2023 onward.

Figure 20
World Annual Change in Production
Vs. Average Price / Barrel of Oil
Political Crisis

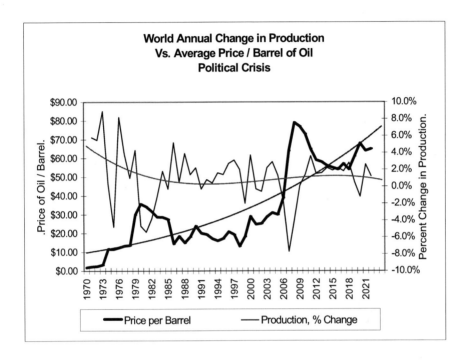

Oil shortages supercharge unsatisfied demand and prices quickly rise to find a new equilibrium. As shown by Figure 21, continuing unsatisfied demand sustains higher prices throughout the Forecast Period.

Oil, Jihad and Destiny   95

Figure 21
World Oil Market: Unsatisfied Demand Vs. Price
Political Crisis

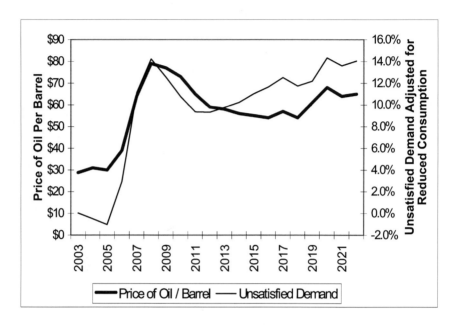

The rate of Inflation and changes in the price for a barrel of oil tend to mimic each other (Figure 22). The "oil depression" causes a short, but brutal spike in the rate of inflation as the price for every product that uses oil as a feedstock escalates and consumer panic causes hoarding. From 1983 - 2002, annual inflation averaged 3.14 percent in the United States. The oil depression characterized by our Political Crisis Scenario increases inflation during the Forecast Period to an average annual rate of 5.46 percent.

By 2009, however, the inflationary impact of oil shortages are offset by the deflationary forces of unemployment, a collapse of world financial markets and a sharp drop in world GDP. The results are shown in figure 22 for the United States.

Figure 22
World Price of Oil Vs. U. S. A. Rate of Inflation
Political Crisis

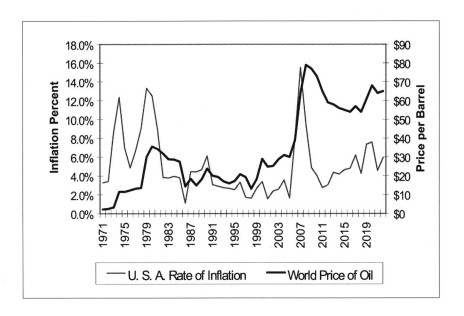

The rate of change for GDP in the United States continues the decline that started in the 1970s. The "oil depression" operates to collapse GDP in much the same manner that the "great Depression" did in the 1930's. As shown in Figure 23, it drops 7.3 percent in 2007 and another 3.1 percent in 2008. Continuing deterioration pushes it negative again in 2020 to minus .15 percent. From 1983 to 2002, the average annual growth of GDP in the United States averaged 6.01 percent. During the Forecast Period, the oil depression drops this average to 2.67 percent. With the conversion of the world oil market from a consumer demand driven market to a producer controlled market, we can anticipate that GDP performance in the United States will be closely tied to the availability of oil.

Figure 23
Rate of Change
N. A. Oil Consumption and U.S.A. GDP
Political Crisis

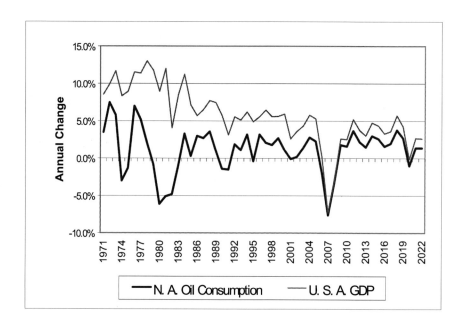

Given the assumptions we used to construct a Political Crisis scenario, unemployment in the United States, which averaged 6.02 percent in the 1983 - 2002 timeframe, jumps to an average of 12 percent in the Forecast Period. It reaches a high of 22.9 percent in 2008 (Figure 24) and never declines below 8 percent thereafter.

Figure 24
U. S. A. Unemployment and GDP
Political Crisis

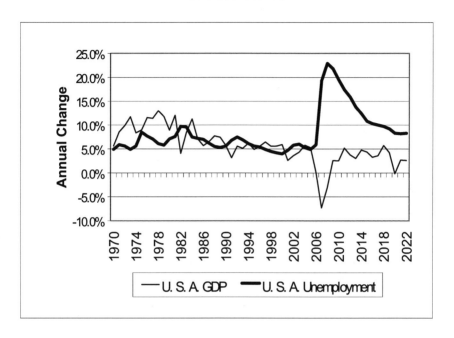

There is, of course, something even more sinister in the data that was generated by the underlying hypothesis for this scenario.

If you live in Western Europe or the Pacific Rim - take another look at the Charts presented for the Political Crisis Scenario. This could be your future.

The sudden reduction of production would force an equally rapid decrease in consumption. The resulting oil shock would cause an uncontrollable economic panic and extreme social turmoil in every industrialized nation.

This prompts a question: How many governments would be able to survive the subsequent political confusion?

Conclusion: If the industrialized nations continue to bicker over their respective political and economic ambitions, if bureaucratic pomposity and obstinate behavior continues to be the hallmark of international relations, and if the United Nations fails to bring forth a consensus solution to the rising energy of radical Islamist behavior, then this scenario becomes a realistic possibility. How we humans handle our military, diplomatic and cultural challenges will decide the course of our economic future. The cultural and economic impact of the Political Crisis scenario is incredibly bad for civilization.

If we fail to deal with Islamist cultural challenge, in Iraq, Saudi Arabia and elsewhere, then there is a 75 percent chance that this scenario - in some form or other - will become reality.

Chapter 7   CHOOSE ANY SCENARIO

There is no simple solution. We can not escape the inevitable.

The Assumptions Are In Line

The production assumptions used to generate the Best Case, Production Crisis and Political Crisis scenarios are in line with other forecasts of oil depletion. As shown in Table 7, all regions - except North America - are expected to produce <u>more</u> oil during the forecast period than they did in the 1983 - 2002 timeframe.

Table 7
Comparison of Scenario Assumptions

	Historical 1983 - 2002	Forecast 2003 - 2022		
		Best Case Scenario	Production Crisis	Political Crisis
World - Compound Average Growth Rate in Production	1.4%	1.5%	.8%	.7%
North America	-.2%	-2.0%	-2.3%	-1.5%
Middle East	3.1%	2.1%	.4%	-.1%
EurAsia	0%	2.2%	1.8%	1.4%
ROW	2.7%	2.0%	1.8%	1.7%
Total Production				
North America	104.1 Bbl	87.9 Bbl	85.8 Bbl	89.9 Bbl
Middle East	130.7 Bbl	205.4 Bbl	175.5 Bbl	135.0 Bbl
EurAsia	112.2 Bbl	156.9 Bbl	153.8 Bbl	142.3 Bbl
ROW	137.3 Bbl	205.9 Bbl	201.7 Bbl	194.5 Bbl
Source: Depletion Model, BP historical data				

It Doesn't Make Any Difference

On a long term basis, there is a remarkable similarity between the Best Case, Production Crisis and Political Crisis scenarios described in this report. Average annual GDP for the United States - and every other nation - will decline. As shown in Table 8, both the Production and the Political Crisis scenarios reduce average GDP growth over the 20 year period from 2003 to 2022 by more than 48 percent versus the GDP growth of the period from 1983 to 2002. Even the Best Case Scenario drives down average GDP by 39 percent. The average rate of inflation over the 20 year Forecast period from 2003 to 2022 will be higher, as will the average rate of unemployment. Oil consumption will decline and the price of gasoline will increase. The price of a barrel of oil, which actually declined in 2002 versus 1983, will increase over the Forecast Period, more than doubling by 2022.

Sure. GDP, inflation and unemployment will vary from nation to nation. But every nation will exhibit similar economic behavior. This is our future.

Table 8
Oil Depletion Scenario Comparison
United States Economic Data

	Forecast Period 2003 - 2022			History
	Best Case Scenario	Production Crisis	Political Crisis	1983 - 2002
Average Annual GDP	3.67 %	3.07 %	2.67 %	6.01 %
Average Annual Inflation	3.21 %	4.40 %	5.46 %	3.14 %
Average Annual Unemployment	6.10 %	6.47 %	12.0 %	6.02 %
Average Annual Change: Consumption	1.52 %	.85 %	.71 %	1.34 %
Gasoline: Average Price per U. S. Gallon	$ 2.41	$ 2.61	$ 3.01	$ 1.21
Oil, CAGR Price per Barrel	4.46 %	5.17 %	4.38 %	- .70 %
Oil, Average Price per Barrel	$ 43.74	$ 47.89	$ 56.89	$ 20.74

In Figure 25, we examine the results of the assumptions made for each scenario on world oil production. Best Case production peaks in 2026 at 37.8 Bbl. In the Production Crisis, oil production peaks at 32.9 Bbl in 2021. The Political Scenario tops out at 32.0 Bbl in 2025. In each case, there is prolonged period of minor changes in production on either side of the peak year.

Figure 25
Scenario Comparison
World Oil Production
1970 - 2042

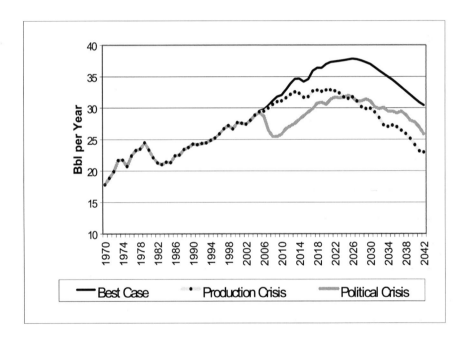

The price of oil will continue to increase. That's the obvious conclusion of all three scenarios. The shortages induced by the Production Crisis push gasoline prices higher than the other two scenarios and by 2022 the average annualized price of gasoline in the United States is almost $4.00 a gallon. Although the panic caused by the Political Crisis causes a spike in gasoline prices, the Best Case and Political Crisis scenarios generate almost identical long term price action (Figure 26). In either of these scenarios, gasoline would cost about $ 3.50 per gallon in 2022.

Figure 26
U. S. A. : Price of Gasoline
Scenario Comparison

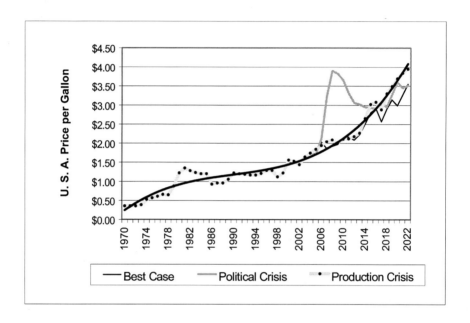

There will be those who believe our gasoline price projections are much too optimistic.

Perhaps they will be right.

Deterioration in the quality of oil feedstocks, combined with petroleum shortages, could force gasoline prices much higher.

Although the 2007 Political Crisis creates a huge spike in inflation, all three scenarios (Figure 27) yield a rate of inflation that ranges between 5 and 6 percent by 2022. By then, oil shortages versus the natural growth of oil consumption will have forced the U. S. economy into an inflationary trend that promises to continually increase the cost of living. We will have a bifurcated economy. On the one hand the price of oil, and any product made from oil (including gasoline, heating oil, propane, lubricants, textiles, asphalt, cosmetics, plastics and chemicals), will become more expensive, increasing the rate of inflation. On the other hand, decreased consumer demand, declining capital spending, growing unemployment, the collapse of the housing market, along with increasing bankruptcies and debt defaults will all act to push down the rate of inflation.

> The industrialized nations
> will begin to resemble closed third world economies.
> High rates of inflation will be accompanied
> by increasing rates of unemployment.

It should be pointed out that in all three scenarios, declining GDP and rising unemployment will inevitably put downward pressure on wage rates in every industrialized nation. The disparity of income between "rich" and "poor" will increase.

Figure 27
Rate of Inflation
United States
Scenario Comparison

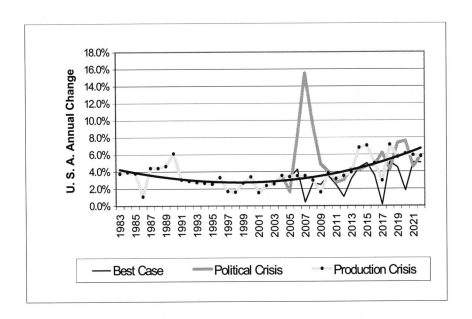

If we compare these three scenarios, we find that relatively small oil shortages can produce a surprisingly strong decrease in regional economic growth. Gross Domestic Product for the United States is characterized for all three scenarios in Figure 28. Aside from the timing and magnitude of the fluctuations, Best Case and Production Crisis scenarios produce almost identical GDP results. In the Political Crisis scenario, the crisis itself sharply reduces oil consumption from 2006 through 2013. This has the effect - ironically - of prolonging positive GDP results because this unused oil is available for consumption through 2022. But in 3 to 5 years, the Political Crisis scenario GDP will also turn negative.

Again, we must note that one could make a chart that looks like this for any industrialized nation. No nation will be immune from the ravages of an oil depletion crisis. It should also be noted that a gradual decline in the average rate of change for American GDP had already started before 2003. Because of the oil depletion crisis, this decline also becomes a long term characteristic of the world's economic structure.

Figure 28
Rate of Change
U.S.A. GDP
Scenario Comparison

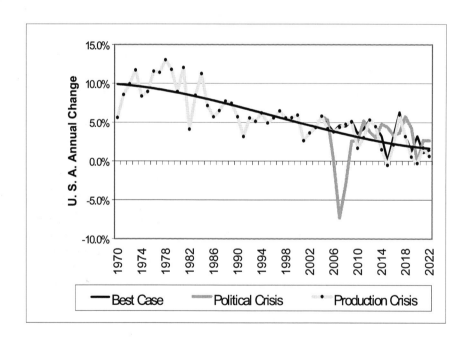

The only real difference between these three scenarios is the characterization of unsatisfied demand (Figure 29). There is little or no unsatisfied demand in the Best Case Scenario until 2014. Then it begins to increase as oil shortages become more common in the world market. The shortages of the Production Crisis start early and climb steadily throughout the Forecast Period. The Political Crisis sets off a surge of unsatisfied demand and it stays high thereafter.

Figure 29
World Unsatisfied Demand
Adjusted for Reduced Consumption

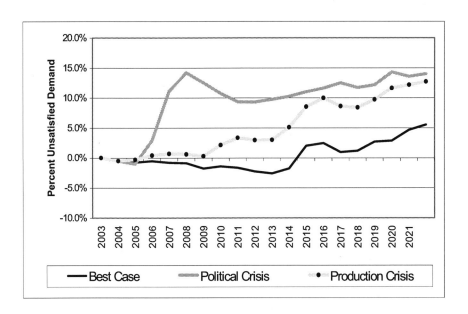

Oil shortages translate into reduced consumption. Nowhere is this more evident than in the reduced availability of gasoline (Figure 30). Although higher prices and unemployment will reduce consumption, the shortfall will have a devastating impact on the cultural lifestyles of the industrialized nations. The Best Case scenario assumptions produce only a modest shortfall of 39.2 billion gallons. Not too bad. However, the Production Crisis assumptions trigger a recession shortage of 715.1 billion gallons, and the Political Crisis restricts gasoline consumption by a depression shocking 1.286 trillion gallons.

Figure 30
World Gasoline Shortfall
Adjusted for Reduced Consumption

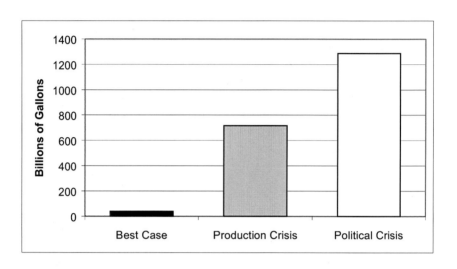

Greater Precision Does Not Help

Many economists will scoff at this analysis. The data limitations discussed earlier in this report prevent absolute accuracy. The scenarios are the result of deduction. The statistical credibility of the projections could be improved. Yes, there are limitations to research based on soft data. Hopefully, someone with a very fast computer and a thoroughly tested econometric model will step forward to help improve the accuracy of my analysis and the viability of my scenarios.

But will greater precision make any real difference to the inevitable outcome?

No.

I ran seven different models. Each characterized a unique variation of production and consumption. The results were painfully obvious. The only way to improve our outlook is to guarantee that two things will happen:

- we humans find another trillion barrels of conventional crude oil by 2022; and
- we find an effective way to deal with the Islamist challenge.

Does anyone really believe that BOTH of these requirements will be satisfied?

Wishful think doesn't cut it. Academic condescension will not be helpful. The pomposity of ideological conviction can not provide a solution.

We humans are stuck with an imprecise future.

And Now: The Good News

All three of the scenarios discussed in this report have assumed that oil market fluctuations and shortages will not cause a corresponding financial crisis of the magnitude that hit the world in 1929.

That's not a good assumption.

America has developed an irresistible affection for debt. From a base of $1.4 Trillion in 1970, America's debt outstanding jumped 178 percent between 1970 and 1980 to $3.9 Trillion. By 1990, America had 7.6 times more debt than it had in 1970, and by 1990 it had 12.8 times more debt than it had in 1970. Debt has continued to pile up. By the end of 2003, total debt obligations exceeded $22.9 Trillion. Intragovernment obligations (including Social Security debt) exceeded $ 3 Trillion. Because of its current-account deficit, the United States had to borrow $1.5 billion from foreign capital markets each and every business day by the end of 2003.

In many ways, the burden of these debt obligations was caused by the relentless pursuit of the American dream. From the 1940s to the 1970s, one income was usually sufficient to support the lifestyle of a typical family. From the 1970s to the 1980s, it took more than one income to sustain the family as the expenses of the desired lifestyle increased. From the 1980s until the 2000s, it took the income of two wage earners plus an accumulation of debt to sustain the American dream.

While family debt was increasing, so was the debt of the government. Political expediency responded to increasing demands for government services and welfare support. Entitlement programs increased rapidly in scope and cost as politicians used tax dollars to buy votes.

Unfortunately, a debt crisis could be triggered by any one of the three depletion scenarios discussed above - Best, Production or Political. At first, ballooning public and private debt, along with a soaring the trade deficit, will sharply depress the dollar, drive up inflation, and reduce the ability of the United States to float additional public or private debt. There will be insufficient capital or income to fund the aggregate debt load of government, business and consumer obligations. Rising defaults will severely decrease the income and wealth of both domestic and foreign debt holders. This inevitable collapse of international debt will put additional downward pressure on the GDP and employment of all debt holder nations. In other words, the collapse of the American economy will bring down the economy of every nation on this planet.

The three Government Sponsored Enterprises - FNMA (Federal National Mortgage Association), GNMA (Government National Mortgage Association) and FHLMC (Federal Home Loan Mortgage Corporation) - collectively finance most of the real estate market's residential component. They are highly leveraged and are holding over $3.6 Trillion in debt obligations. Much of this debt has been turned into collateral notes called a Mortgage Backed Securities (MBS) for sale to banks, mutual funds, pension funds, insurance companies and the public. Unfortunately, these securities act like bonds. If interest rates go up, the value of the MBS goes down. The financial stress of an oil crisis could easily force interest rates up if buyers became afraid of the liquidity value of America's debt obligations. All hell would break loose. The collapse of derivative market liquidity, real estate loan defaults, declining mortgage and mortgage backed security values, and declining cash flows would force the GSEs into bankruptcy. If confronted by the imminent financial collapse of one or all of these housing GSEs, Congress would be forced to bail them out. But where will Congress get the money? From a deteriorating tax base? By printing money? How could Congress borrow the money to bail out the GSEs if the financial system is imploding? Will the term "full faith and credit of the United States Government" be sufficient to secure an astronomical increase in public debt?

And what about consumer and corporate debt burdens? Either the Production or Political scenarios could easily trigger a debt crisis as thousands of loans go into default. If the financial system collapses from excessive debt, then unemployment will be much higher and last far longer than discussed in our scenario characterization. A financial crisis would also place far more downward pressure on GDP - just as it did in 1929.

Of the three scenarios, the oil shock of the Political Crisis would cause the worst financial disaster. The sudden jump in oil prices and the costs associated with oil shortages would trigger a short spike in the rate of inflation. The collapse of the financial markets, declining real property values, soaring unemployment, decreased consumption and failing capital investment would then propel the world economy into a brutal period of deflation.

Although the Production Crisis would produce similar economic results, the process would be more gradual, giving national governments more time to adjust monetary policy and debt obligations. The net impact is more likely to send the world economy into a short period of deflation followed by an extended period of inflation accompanied by trade wars, increasing unemployment and declining GDP.

> But the financial crisis will not be limited to the United States.
> Given the parameters of the Production and Political Crisis scenarios,
> it is highly likely that by 2022,
> no government will have the financial resources
> to sustain its social welfare programs.

A financial crisis within the parameters of the Best Case Scenario is more likely to look like the recession of 2000, followed by a prolonged period of economic decline.

In other words, if an oil crisis is accompanied by a financial panic, then the assumptions of all three scenarios discussed in this Chapter could be overly optimistic.

Chapter 8 CONCLUSIONS

Truth is unpopular. In order to avoid the economic and cultural ravages of depletion, we will have to impose restrictions and changes on human behavior that most people will not like. Too bad. If we fail in our quest for solutions, we risk the distinct possibility that our destiny will be determined by the four plagues of the 21st century.

Summation

It's incredible.

We humans inherited at least 3 Tbl. of oil. Or 2.6 Tbl of oil. Or 3.9 Tbl of oil.

Pick your expert. Everybody has a number.

And since we have only used 890 Bbl, more or less, we must have a lot of oil left. Feel better?

Hmm. Perhaps we should look at the composition of our windfall.

We will never be able to find, extract and refine
all of the oil that is left on this planet.

In order to use all of our oil resources, we humans would have to find and liberate every drop of oil on this planet, ignore all of the environmental consequences, and be able to pay an unlimited amount of money for the resulting oil based products.

That will not happen.

We also have to make the absurd assumption that the amount of oil energy we get from all of this effort will be greater than the energy it will take to extract and process the oil we find.

That will definitely not happen.

There will be multiple technical challenges. Something like 350 to 530 Bbl of our legacy is buried in the ground as unconventional heavy oil, bitumen and shale oil. Getting this sticky stuff out of the ground will take enormous amounts of energy and water resources. As for the 1.4 to 2.1 Tbl of (increasingly sour) conventional crude oil, and 300 Bbl of NGL, it will not do us much good unless we find all of it. Then we will have to extract, transport and refine this stuff using enhanced recovery technology and increasingly expensive refinery processes.

Human endurance and ingenuity will be tested over and over again. Oil exploration, production and transportation activity requires the pursuit of projects that are located in an increasingly hostile working environment - under the ocean, under layers of ice, in remote regions, and so on.

Human conflict will restrict oil exploration and production activity. Oil projects will collide with cultural antipathy and labor discord. We can not assume that the use of military force will always be a feasible solution, nor can we assume that throwing money at conflict will bring about the desired result.

The cost of finding, producing, refining and distributing oil
will exceed the price that we humans are able to pay
for oil based products
long before we run out of oil.

We can't build a credible production versus consumption model on a pure speculation that defies current discovery trends. So we do not. But in the final analysis, it doesn't matter how much oil there is in the ground.

What really matters? How much of this stuff can we produce?

And that forces us to consider all kinds of variables - geography, geology, exploration technology, extraction, transportation, refining, weather, cost versus price constraints, the economics of consumption and the biggest wild card of all - culture.

We have to face reality.
Most of the world's oil is under the feet of a culture
that has serious doubts about the industrialized world.

Distrust. Aversion. Occasional hatred. We need to deal with this reality in making our assumptions about production.

And then construct scenarios that reflect our analysis.

This is what we have done. Three scenarios. The Best Case Scenario assumes cultural conflict will have little impact on the availability of oil. Instead, it tries to project a reasonable assessment of current trends. The Production Crisis portrays a growing conviction that oil production volumes should be established to meet the needs of the producer - not the demands of the consumer. The Political Crisis scenario characterizes what would happen if cultural conflict erupts into confrontation.

In all three scenarios, oil production peaks somewhere between 2021 and 2026 at 32.0 to 37.8 Bbl per year. That may sound like a lot of oil, but it is not enough.

There is not enough oil in any of these scenarios
to sustain the world's economy.
Recession is inevitable. Depression is possible.

Let's review our conclusions.

The Best Case Scenario: In order for us to experience the relatively mild economic decline of this scenario, average annual world oil production would have to grow throughout the Forecast Period. We have to be able to find - and produce without disruption - all of the oil it will take to sustain a modest increase in world oil consumption. There would have to be an effective spirit of cooperation between producer and consumer nations. Everything has to work. No political posturing. No bureaucratic mistakes. No under the table deals for extra oil. And perhaps most important of all - No sustained threat of extremist disruption in either the oil producing or consuming regions.

Are all these assumptions possible? Probably not. The probability that the Best Case Scenario presents an accurate projection of the future is less than 40 percent.

The Production Crisis: This study makes a very reasonable set of assumptions. OAPEC production capacity stalls at a maximum of 25.4 million barrels of oil per day. Although there is a minimum of armed conflict, cultural constraints limit the use of foreign technicians and engineers to sustain production. Exploration is lethargic. Depletion gradually restricts production. Output is erratic and demand exceeds supply. We humans fail to solve our planet's energy problems. If these assumptions are correct, (or even reasonably correct), then this scenario is the alternative to the Best Case Scenario with a 70 percent probability. Further, if we fail to work together to solve our planet's energy problems, then international conflict is inevitable as a Production Crisis tightens its noose around the windpipe of economic growth.

If armed conflict happens - the economic characterization of the Production Crisis is much too optimistic.

The Political Crisis: A wild card. If the industrialized nations of the world continue to bicker over their respective political and economic ambitions, if bureaucratic pomposity and obstinate behavior continues to be the hallmark of international relations, and if the United Nations fails to bring forth a consensus solution to the rising energy of radical Islamist behavior, then this scenario becomes a realistic possibility.

114

> How we humans handle our military, diplomatic and cultural challenges will decide the course of our economic future.

Within a few years the OPEC nations -- or whoever controls them -- will be in effective control of the world oil economy, and, in essence, of human civilization as a whole. If the industrialized nations vacate Iraq before that country has a self sustaining government, and/or there is a change of government in Saudi Arabia, then there is a 75 percent chance that the Political Crisis - along with the depression it triggers - will become our future.

We can draw some other conclusions from the production, consumption and economic impact analysis of these scenarios:

1. We are running out of oil. Even if we find another trillion barrels, that discovery will only serve to delay the inevitable oil crisis.

2. Our immediate problem is not - How much oil is there in the ground? The real question is - How much oil can we produce?

3. At the point of peak production, there will still be plenty of oil in the ground. It will just get harder and more expensive to find, produce and refine it.

4. As we approach the peak of oil production, we will have periods of surplus that alternate with periods of deprivation. These cycles will become extremely volatile.

5. Each cycle of shortage will force a corresponding decrease in consumption. The economic impact will be recessionary. When there is a subsequent surplus, the world economy will slowly recover.

6. The oil market has transitioned from a consumer driven market to a producer controlled market. Future consumption will be limited by production.

7. We are moving from a world economy that enjoyed excess oil capacity to a world economy dominated by chronic, severe, and highly volatile shortages. The GDP of all nations must respond to this volatility. Shortages, whether periodic or sustained, will drive GDP lower. This is true for all three scenarios. Absent alternative sources of energy, world GDP must inevitably decline and may turn negative before 2022. The social and political cost will be enormous.

8. If there is an oil shortage induced financial crisis (likely), then there will be a period of severe deflation. Otherwise, the long

term net impact of an oil depletion crisis is inflationary. By 2022, the average annual rate of inflation - even with the deflationary impact of higher unemployment and financial chaos - may exceed 5.5 percent. It will continue to increase.

9. Unemployment will go up. Although the timing and rate of increase will be determined by the characteristics of the eventual oil depletion scenario, by 2022 it could easily exceed 7 percent in the United States.

10. Although periods of surplus capacity will cause temporary reductions, average oil and gasoline prices will more than double by 2022. Price acceleration has already begun.

11. Oil shortages will force consumers in all of the industrialized nations to change their lifestyle before 2022. They will use less oil based products - including heating oil and gasoline.

12. Unless we develop and deploy a new energy system over the next 10 to 15 years, we will begin to burn coal, wood, trash, and animal waste for cooking, heat and power. Vehicles will use increasing amounts of fuels derived from coal gasification and plant materials. Air and water pollution will increase.

13. There aren't any conservation measures that could conserve enough energy to make up the difference between demand and supply. For that matter, there are no renewable energy projects on the horizon that could help us to avoid an energy crisis. Our need for energy far exceeds any relief we could achieve from these measures.

14. The financial chaos of a Production or Political Crisis scenario will not be limited to the United States. It is highly likely that by 2022, no government will have the financial resources to sustain its social welfare programs.

15. By 2022 we are expected to have more than 8 billion inhabitants on this planet. We can not support them with dwindling resources of energy and we are not prepared for the consequences.

Competition and Conflict

We will compete - nation against nation - for oil. This competition is already underway.

- Why would China have a diplomatic and military interest in the Middle East? Oil. By 2015 they will need to import over 8 Mbl a day. China has been willing to trade access to oil for weapons of war. Will Muslim leaders favor contracts with non-Jewish, non-Christian nations?

- Why would Russia care about the Caspian? Oil and money. Russia needs the revenue and independence that oil can bring. Oil means military strength, gasoline for vehicles and fuel for residential heat. Oil can be used to gain political importance over nations that must struggle for resources.

- Why does the United States care about Iraq? Oil. The U.S. imports over 50 percent of its oil and this percentage is increasing year by year. By installing a friendly regime in Iraq, the United States becomes a favored customer of that nation's rich reserves of oil.
- Why did France and Germany oppose the U. S. action in Iraq? Oil. Money. Both nations had contracts for oil and equipment with the old regime. Both nations need access to a stable supply of oil because there is little left in Western Europe.

Oil is a reality that exceeds the importance of ideology, morality or politics.

Except for religious fanatics. For them, oil is a weapon of war. Terrorists with ties to al Qaeda, Hizballah, PFLP, Abunidial, Islamic Jihad, Muslim Brotherhood, Pasdaran and other Muslim paramilitary political groups will disrupt the flow of oil to the West whenever they can. It will become a basic part of their strategy to bring down the West and current events suggest they may succeed. If so, we can accelerate the Political and Production Crisis timetables described in this report. Oil depletion is upon us.

And can we protect our oil supplies from terrorist activity? No. Some estimates put the military costs of protecting pipelines and tanker routes from terrorist activity, primarily paid by U.S. taxpayers, at around $15-20 a barrel. If that military action becomes a necessity, then we can sharply increase our estimates of inflation, the price of oil and gasoline, and the economic devastation of oil shortages.

Yes, this international competition is all about oil. And the competition among consumer nations will become much more heated as our combined desperation mounts. We all share a common crisis.

Oil depletion.

As shown in the Production Crisis scenario, the world will enter a period of permanent recession with unemployment rates of more than 7 percent. For some nations this will translate into a permanent depression with unemployment rates of more than 15 percent. Unless alternative energy resources are developed, economic activity must contract to match the available supply of oil.

Oil consumption will be forced down by shortages.

Oil demand will be forced down by increasing prices.

Oil will eventually become too valuable to use for the production of gasoline.

Conservation will be enforced by the police power of the State and the hard realities of economics. The economic impact of constrained consumption will be devastating.

> Every industry and every job that depends on oil based human mobility will be in jeopardy.

There will be thousands of bankruptcies. The Best Case, Production Crisis and Political Crisis scenarios all produce a stunning impact on Western Civilization. The differences between them have more to do with the timing of the resulting economic distortion than of long term substance.

For example, let's assume that the world's natural growth in oil demand from 2003 through 2042 averages 1.8 percent per year over this 40 year period. There will be a growing delta between this natural growth in the demand for oil and the restrictions imposed on the consumption of oil by future production shortages. Using Best Case scenario data, we can calculate an annual natural demand growth of 1.8 percent, then adjust this assumption for the impact of higher prices, substitution and conservation, and plot the resulting net unmet demand versus available oil for consumption. If there were no production restrictions, world economic growth would generate a natural demand for 1.628 Tbl of oil over this 40 year period. Because oil production is declining, however, users would be forced to reduce their consumption to 1.397 Tbl. The net delta difference of the resulting shortage exceeds 230.8 Bbl of oil and 4.5 trillion gallons of gasoline.

At some point, there will be a stock market crash in every nation that harbors financial investment exchanges. Financial institutions will be devastated by mounting uncollectable personal and corporate debt instruments. As the oil crisis progressively disrupts economic activity, corporate cash flow will gradually be constrained by decreasing revenues and uncollectable debt.

Government will save us. Right?

Probably not. Look at any poor nation. Really poor nation. Disease, starvation, poverty, crime, conflict - they are a way of life. The government does not have enough money for social programs like education, health care, food, shelter and clothing.

> How can any nation continue to support the social welfare of its citizens if tax revenues are tied to a declining GDP?

So we can anticipate a spate of temporary social welfare programs for the unemployed and the victims of a declining economy. But when

118

the money runs out, when government is unable to float any more debt, then there will be a severe and permanent reduction in social welfare spending. National governments will be unable to fund even the most basic of services - road maintenance, water systems, schools, hospitals - and so on. There will be a steady deterioration of each nation's infrastructure.

No. Government can not save us. No government or collection of cooperating governments will be able to stop the oil depletion crisis from happening. They will not even be able to manage the crisis, or buffer the population from the crisis. There will be charges of favoritism (against those who appear to have more access to oil products), charges of collusion with foreign governments and big oil companies, and charges that the national government has failed to protect the selfish best interests of its citizens.

> Most national governments will fail to understand the oil crisis,
> prepare for the oil crisis
> or even develop a sane policy to mitigate its effects.
>
> Instead, there will be confusion and lies.

It will be business as usual. Imperious, self-serving, politics. The desire for political power disdains the use of long range planning that might annoy the masses.

There will be a lasting frustration with the economic downturn. A frustration with heating oil and gasoline shortages. A frustration with oil product price inflation. A feeling of helplessness. Then anger. Strikes. Riots. And bloodshed. Civil and national conflict. War is a catharsis for anger. The four plagues of the 21st century - disease, famine, violence and lethal misery - will bring death to millions.

The oil depletion crisis will inevitably lead to the destabilization of national governments. An us Vs. them mentality develops. Us versus the oil producer nations. Us versus other consumer nations. Us versus the terrorists. Us versus anyone who disrupts our oil supply. Liberals versus conservatives. Muslims versus Jews and Christians. Environmentalists versus everyone else.

Find someone to blame. And then hurt them.

Freedom of movement and independence will disappear. The hungry and the cold will take whatever steps are necessary to survive. Burn anything - coal, wood, tires, garbage, animal wastes - burn anything that will produce heat for warmth and cooking. Morality and ideals will lose their value. Steal what you need. People will forget how to rely on themselves as individuals. They will attribute their status to "destiny", "the state", "evil politics" and "mysterious forces". They thus become "the masses", ripe for revolutionary leadership.

Alternative voices will rise to fill the air with solutions. Mostly despotic. Dictatorships can - after all - make things happen. We humans will realize that our current national governments are powerless, confused and contentious. We tire of left versus right. A charismatic extremist will arise, promising to make things better, to relieve our frustration and calm our anger. The new order will heat our homes, feed our children and give us back our cherished personal vehicle mobility.

Democracy teeters on the edge of oblivion.
Desperation creates opportunity for new ideologies.

The flow and price of oil increasingly depends on cooperative planning between all nations.

Or there will be conflict. Economic depression.

Do we really believe such cooperation is possible?

Or will humanity slip backward into a 21st century version of the dark ages?

Bearish or Bullish?

This report will be criticized. Some will claim it is too bearish.

They may be right.

In 1959, Weeks published a report that claimed we had 2 Tbl of oil left on this planet. On at least two dozen occasions since then, other researchers have also claimed that we humans have 2 Tbl of oil (plus or minus 500 Bbl) left on this planet. So for almost 45 years we have been doing these oil depletion studies and for almost 45 years we have been coming to the same conclusion. And then by golly, we seem to be able to find enough new oil to fulfill our projection.

So, we can assume that this state of affairs will continue forever - right?

Wrong.

The facts work against making this assumption. The research that went into the No Change, and Best Case scenarios show that oil depletion now looms as a definite problem that we humans will have to resolve.

- We humans have been consuming oil faster than we have been finding it since the early 1980s. We can not expect to produce more oil unless we find it. Oil is being discovered at the rate of approximately 6 billion barrels per year. We are consuming it at an average rate of 28 billion barrels per year. That means that over 75 percent of the oil we consume each year is not being offset by discoveries of new oil reserves. Every year of production creates an incremental deficit.

- North America used to be a major oil producer. That has changed. Oil production peaked in 1970. It has been declining - on average - ever since.

- It is likely that all of the other oil producing nations will reach their respective peak production before 2026. This includes Saudi Arabia and Iraq, where cultural problems - not the availability of reserves - will curtail production.

- In recent years, a number of the countries in the oil producing regions have experienced varying degrees of one or more of the following: economic instability, civil war, political volatility, violent conflict and social unrest. Oil production is vulnerable to disruption in these regions.

- The market for oil has changed from a market driven by consumer demand to a market driven by the availability of production. A chronic surplus of oil has been replaced by an increased vulnerability to shortages and higher prices.

- There is a continuing risk that incremental additions to the production of oil will not occur in time to offset the corresponding growth in demand. Further more, over 250 Bbl of claimed reserves may not even exist.

This report will be criticized. Some will claim it is too optimistic. They may be right.

The Best Case, Political Crisis and Production crisis scenarios all assume we have 877.4 Bbl of "Proven Reserves", we can find and exploit 291.7 Bbl of new reserves, achieve a healthy 266.5 Bbl increase in reserve growth and produce 740.0 Bbl of unconventional oil. If these projections fail to materialize, the oil depletion crisis will be with us much sooner than described in this report.

Only one thing is certain. Oil depletion is a dreadful reality that we humans will have to resolve.

Or mother nature will do it for us.

Nature abhors a vacuum or an imbalance. If our planet is overpopulated, nature will restore a balance through disease and starvation. Man will augment this process by engaging in armed conflict for the possession of dwindling resources. Individuals will suffer the pain of lethal misery. There will be a growing divide between nations that make an attempt to control population growth and those that can not control personal virility. Wealth among nations will become a

Oil, Jihad and Destiny

relative attribute - prosperity simply means that a few cultures are in a better position to survive. The general population will nevertheless suffer.

Civilization must regress. Famine, disease, ignorance, envy, and want banish reason. Survival demands that the deprivations of the oil crisis must be overcome by any means - including self-righteous hatred for "them".

Is there any defense against the four plagues of the 21st century?

Are disease, famine, violence and lethal misery our inevitable future?

Chapter 9   RECOMMENDATIONS

We have reached a fork in the road of human existence. We can create our future or let misfortune rule our destiny. Which road will we take?

Mitigation

If we take a competitive approach to oil depletion that is embedded with deceit and contemptuous behavior, then the awful projections of suffering described above are a given. If we want to avoid competitive oil resource conflicts, we must develop an international oil depletion management plan. We must try to develop a consensus based strategy for sharing our planet's dwindling hydrocarbon resources. Producer nations must be encouraged to cooperate with consumer nations in the creation of agreements for hydrocarbon infrastructure investment, production and consumption. Understandably, this will be difficult because all nations will struggle to protect their selfish best interests. But we have to try.

At the same time, the leading industrialized nations must work together on the development, manufacture and deployment of an alternative energy system for both mobile and fixed site applications.

We can not escape the negative impact of oil depletion.
We can, however, mitigate is effects.

Thus we can put together a list of recommendations. Some are easy. Some are controversial. Many challenge cherished ideology and preconceived conviction. All are worthy of intelligent, thoughtful, discussion.

Identify Available Oil Reserves

We need better information. We must identify how much oil is really left on this planet. From this base, we can then develop a forward plan for production and consumption as well as a timeline for the development of alternative fuel systems. We therefore need to bring together an independent Energy Information Analysis group, drawn from the leading consumer nations and funded by their respective governments. The key word is "independent". The group must not have a political or economic agenda. The staff must include geologists, exploration and production experts, refinery engineers and business development personnel. Its job will be to give us a credible evaluation of available oil reserve, undiscovered reserve and reserve growth data. The examination will include both conventional and unconventional resources. It may be necessary to give this group the power to subpoena private records. But their task is to answer a simple question. How much oil will we humans have available for consumption over the next 40 years?

We also need to drill holes in the ground. The USGS has identified multiple geologic zones that may (or may not) contain oil bearing strata. We need to identify these formations and drill test wells to confirm whether or not oil really exists in these areas. Since these test wells are highly speculative, the drilling costs should be underwritten by the consumer nations that will benefit from the incremental production. Discoveries of oil should lead to contracts between local national agencies (where the oil has been located) and the companies that will extract the oil. The cost of the test wells would be reimbursed from the proceeds of production.

According to some estimates, there is over 1.6 Tbl of oil locked up in oil sands and shales. Even the most optimistic geologists believe that only a fraction of that amount is recoverable. Thus, we need to make a realistic assessment of how much of this oil can be recovered as various levels of cost. We also know that at some point, the energy we produce from sands and shales will be less than the energy we burn in the production process, rendering further production useless. Where is that point? And finally, there is a very real question of where we will get the energy needed to heat the oil in these sands and shales so that it liquefies enough to flow into the recovery process. What are these energy constraints and how will they impact the total volume of reserves?

Result: This process will maximize our knowledge of available oil resources and provide a basis for planning an orderly transition from oil to alternative energy systems.

Reduce Oil Consumption

Develop and enforce cooperative production agreements.

When we consume commodities - such as coal, oil and natural gas - our demand is typically moderated by the price of the commodity we are purchasing.

In an open market economy, the pricing mechanism will typically force consumption to match production. If there has been excess production, lower prices will decrease further production and may increase demand to the point where increased consumption absorbs the available production. However, if natural demand exceeds production, then higher prices will typically drive down real demand until there is a balance between consumption and production. In both cases, the pricing mechanism has adjusted demand to bring consumption in line with production.

Market pricing has the advantage of being the easiest way to adjust consumer demand. Individual consumers make their own lifestyle adjustments - they drive less and buy fuel efficient vehicles. Coal replaces oil in some energy applications. Alternative fuels and energy systems receive more attention. Unfortunately, in our depletion scenarios, higher oil prices will have a negative impact on the economy of all nations and force consumers to shift their allocation of purchase dollars from non-oil goods and services to those governed by increasing oil prices. The economic impact of this shift in spending tends to be inflationary and potentially the cause of a recession as consumer, government and enterprise spending decreases for non - oil products.

It is also possible to enforce a reduction of demand through the mechanism of structured shortages. Nations use the police power of the State to enforce restrictions on consumption. It's called rationing. There is an artificial reduction in demand because consumers are forced to buy less. Demand is adjusted downward by rationing until consumption equals production. The advantage of this tactic is that prices can be rationalized to fixed levels. Consumers receive less product - gasoline for example - but the gasoline they do purchase costs less than it would if the market pricing mechanism were used to force a reduction in demand.

Consumption restrictions also force consumers to drive less and purchase fuel efficient vehicles. Coal, alternative fuels and new energy systems also receive more attention. But rationing has its disadvantages. The relative price stability of consumption restrictions may be attractive, but using the police power of the State to enforce compliance will be cumbersome, expensive, and subject to wholesale cheating. Every special interest group will scramble for additional allocations. Black market racketeering is inevitable. Restrictions become political fodder.

World oil production, up or down, tended to exacerbate the economic volatility of national economies in all three models discussed in this report. Wide swings in the availability of oil were accompanied by adjustments to GNP, the rate of inflation and unemployment. Gasoline prices bounced up or down in response to alternating periods of excess and restricted production. The estimated social, economic and political cost was enormous.

So what do we do?

Match consumption with depletion[32] . If consumer nations can reach long term oil production agreements with producer nations that

---

32 An idea recently proposed by Colin Campbell in a document entitled "The Remini Protocol" at an energy conference in Rimini, Italy.

encourage a gradually decreasing - *but reliable* - flow of oil, and if long term oil allocation is done on an international basis, then the pricing mechanism can be buffered from the shock of radical changes in production. There will be a gradual, but predictable, increase in petroleum prices. It will be easier for national economies to adjust for petroleum price changes through conservation, demand reduction, and product substitution. The negative economic impact can be mitigated.

Result: If consumer nations are willing to acknowledge oil depletion as a fact of life, and if consumer nations are able to join with producer nations in a long term international production agreement that assumes a gradual reduction of consumption, then we have an opportunity to mitigate the economic impact of oil depletion. The pricing mechanism, combined with a gradual decrease in production, will encourage a corresponding decrease in consumption.

Dump CAFÉ.

Experience in the United States has shown that Corporate Average Fuel Economy (CAFE) standards are an ineffective way to reduce gasoline consumption because they attempt to penalize the manufacturer, rather than the consumer, for the sale of low efficiency motor vehicles. CAFE standards do not directly encourage consumers to decrease gasoline consumption. Any effective program to reduce gasoline consumption must continually raise consumer awareness since consumers - in general - are either not aware of their vehicle's fuel efficiency or they care about the vehicle's features more than its fuel efficiency. Until there is obvious pain, consumers are not likely to purchase more fuel efficient vehicles, shift some of their driving to the most efficient car they own, reduce the number of miles they drive (such as by carpooling, using public transportation, or forgoing long trips), maintain their vehicles better, or reduce their speed.

The existing American car (27.5 MPG) and truck (20.7 MPG) standards and the roughly 50/50 mix of current car and truck sales imply a unified standard of 24.1 MPG. We must substantially increase this average if we want to mitigate the pain of an oil shortage.

Theoretically, a ten percent increase in the price of gasoline should produce an immediate 2.7 to 3.2 percent decrease in gasoline consumption. As consumers change their driving habits and their choice of vehicles, the long term reduction from each ten percent increase in the price of gasoline should be in the range of 3.0 to 4.6 percent. Consumers need an incentive to drive less. Otherwise, the availability of fuel efficient vehicles merely encourages them to drive more miles - thus reducing the potential conservation of fuel.

Unfortunately, the elasticity of gasoline consumption decreases as the price increases because the personal hardship of reduced consumption also increases. In other words, people do need personal transportation in order to function in a social structure that favors private versus public transportation. In addition, millions of small businesses must have a vehicle in order to function. These characteristics of elasticity mean that each successive ten percent increase in price will bring about a smaller reduction in consumption. We must also contemplate that higher gasoline prices will decrease the level of economic activity. Gasoline prices must not be increased to the point where they bring about higher unemployment or a possible recession. Never-the-less, we need to evaluate if a 100 percent increase in the cost of gasoline would produce a corresponding 22 percent decrease (or more) in gasoline consumption.

Proposal: gradually increase gasoline taxes over a period of 5 years to a point where the price of a gallon of gas exceeds $2.50 in 2003 dollars. Use 100 percent of the proceeds to fund the development of an alternative energy system and to support the construction and maintenance of our public transportation infrastructure.

Result: As shown by the experience of Western Europe, higher gasoline prices are an effective way to reduce oil consumption and the proceeds will be used to fund the transition to a new energy economy.

Move consumers to Hybrid Vehicles.

The introduction of increasingly fuel efficient hybrid vehicles that use a combination of electric and conventional combustion engine technology can make a very large dent in our consumption of oil. Over time, it is reasonable to believe that vehicle manufacturers will be able to increase the average mileage of a consumer vehicle (automobiles, light trucks and SUVs, etc.) by more than 30 percent. Manufacturers will be able to stress that these "green" machines are comfortable, powerful, vehicles. Truck and van versions will have 110 volt outlets that can be used for power tools and compressors. Once in volume production, Hybrids will only cost from $2500 to $3500 more than comparable conventional vehicles.

It is currently feasible to build a comfortable sedan with the power of a six cylinder engine that gets 40 MPG and an upscale SUV that achieves 30 MPG. Hybrid fuel economy will be attractive for small to mid-size sedans, SUVs and light trucks. For larger and heavier vehicles, the best fuel efficiency solution will probably be a version of the diesel engine that burns fuel made from clean coal technology.

Unfortunately, fuel economy and cleaner air do not sell vehicles to the American consumer. Most believe that gasoline prices will be low enough so that they do not have to change their driving habits or vehicle preferences. Consumers have to believe that gasoline prices will go higher, and stay higher, before they begin to think of fuel economy as a purchase criteria. This has, of course, been the experience in Western Europe where gasoline has been relatively expensive for many years.

Experience shows that consumers will become far more concerned about fuel economy when they have to struggle with gas shortages - as they demonstrated during the oil embargo of 1974. If we are to avert an economic disaster, however, the government has to nudge consumers toward fuel efficient vehicles before a shortage occurs.

Result: If we combine this recommendation with higher fuel prices, oil consumption for vehicle applications can be reduced up to 40 percent over a period of 15 to 20 years. This will delay the depletion of our oil resources and improve our chances to avert a long term oil recession.

Encourage new attitudes about personal transportation.

Ever since the horse became a symbol of independence and freedom, men have clung to the concept that individual freedom equals having a personal means of transportation. In contemporary America and the other industrialized nations of the world, that belief is also widely held by women - including millions of mothers who provide a family taxi service. Moving from personal transportation to public transportation is going to take a massive change in fundamental cultural attitudes.

But we all know, or at least sense, that this business of building more freeways and paving over the good earth with parking lots has to come to an end. Instead of encouraging urban sprawl, we have to think compact communities and public - rather than private - transportation. Paving over farmland creates an energy intensive economy.

That change in cultural attitude starts with making changes to local zoning laws. These community planning rules currently favor urban sprawl by restricting the construction of higher density residential units. This effectively forces families to find a place to live miles away from their employment. This has to stop. Where environmentally feasible, we need to encourage zoning that links residential density to local employment opportunities.

It would also be helpful if it were easier to enjoy at least some of the financial benefits of home ownership while also increasing domicile mobility. Job change is a fact of modern life and needs to be accompanied by an equally easy way to transfer equity to a new residence.

For those that must commute, or have children in local schools, ride sharing becomes an important factor in energy conservation. In times of oil shortages, such as those envisioned in this report, ride sharing will be forced on us. We need to develop a system of casual ride sharing - hitchhiking - that permits those on foot to find a ride. Available communications technology can be used to provide the connection and emerging security technology can make casual ride sharing a relatively safe event.

All of this points to another cultural change. We need to replace a loose collection of people that happen to live in the same neighborhood with a fraternal sense of local community. It is from this foundation that a source of cooperation and trust can emerge. Ride sharing and domicile mobility are an easier in an environment of friendship and mutual support.

Result: Our attitudes about personal transportation will change. Eventually, these attitudes will contribute to greater fuel consumption efficiency.

Force politicians to get serious about public transportation.

When we moved to the foothills, both my wife and I were very pleased to find that we could take Amtrak (our rail passenger service) to the Bay Area. It was great. I could ride down in the morning to San Jose, Santa Clara or San Francisco, arrive relaxed and ready for the challenges of the day, and then in the late afternoon I would take an equally carefree train back home. But government bungling soon got in the way. First, it was decided that passengers could not take the bus to the train station unless they were actually going on the train - thus discouraging those who simply wanted to use public transportation as a way to get into Auburn or Sacramento. Then Southern Pacific, the owner of the tracks, was allowed to give service priority to freight trains, guaranteeing that the passenger trains would frequently be held up by slower moving freight service. Then, bus service replaced trains for portions of the route, making the trip infinitely more tedious.

I gave up. Public transportation was no longer a feasible alternative for me.

Too bad.

What this shows, however, is that our government officials and pedantic bureaucracy are only giving lip service to public transportation.

This has to change. Forcing politicians to support public transportation must be a top election issue.

Result: If we succeed, then consumers will be able to migrate from personal to public transportation. It must be available or it will not be used.

Change the Railroad business model.

My wife and I often hike in the Sierras. One of our favorite treks takes us up a mountainside to a lovely group of small lakes where we stop for a welcome rest and a light lunch. The trial passes over the Southern Pacific Railroad tracks which laboriously climb up the side of the mountain to Donner Summit. On our last climb of the season last fall, we stopped to rest alongside the tracks. It was very quiet. Not a train in sight. The only noise was the roar of the truck traffic on highway 80 below. I turned to my wife and said: "What's wrong with this picture?"

The problem, of course, is that while truck after truck makes its way between California and Nevada, only a few trains are using the tracks which parallel the highway. As an economist, it is painfully evident our railroad tracks and right-of-way are one of the most underutilized capital assets in America.

The railroad business model was created when entrepreneurs were the only ones able to raise enough capital to build the rails and buy the rolling stock. Governments frequently subsidized these ventures with land grants and other incentives. One hundred years ago, that was a good business model.

But it ain't any more. Rail right-of-way and capital improvements are far to valuable to be monopolized by one company. We need a new business model. One that converts rail links into publicly owned steel highways for <u>commercial</u> traffic. A national rail system augmented by State and regional rail links.

Passengers and freight.

Result: New entrepreneurs will figure out how to combine highway and rail into a more fuel efficient transportation business model.

Rewire America (and other countries).

This is one for the environmentalists.

One of the easiest ways to reduce fuel consumption is to let knowledge workers "telecommute" from their home to their office, students to engage in distance learning, and consumers to shop on-line. Unfortunately, safe, secure and private broadband service to the home is NOT generally available. For various technical and economic reasons, neither cable nor satellite service is a long term solution. Wireless technology offers another set of frustrating problems. The telephone companies have strangled DSL. Because of this, the credibility and continuing use of the Internet for essential communication services is at risk. Most of the proposed interactive services will fail to reach their best audience. For the poor and the remote, multimedia transmission is a joke.

As a nation, we have to decide what we want to accomplish. Every home, every school, every institution and every Enterprise must have functional broadband access to multimedia services. We have to rewire America. We have to replace every copper connection with a fiber optic link. Right to the PC, appliance and media devices we propose to use for communication. Yes. This means replacing all of the local loops. And all of the switching system equipment that makes high speed transmission possible. And all of the software needed to make this system secure.

This is an enormous infrastructure project. It will take years of dedicated service by a very large company to accomplish. Capital expenditures in the billions of dollars will have to be financed through earnings growth and the creation of gilt edged debt. In terms of organizational structure, this effort is analogous to the one our nation embraced when our government endorsed the concept of universal telephone service over 60 years ago. Back then, universal service seemed impossible and overly expensive. But we found a way to achieve our objectives.

Unfortunately, the administrative structure to make this happen does not exist. It was destroyed when the Federal Government broke up the Bell System. People who represented themselves as consumer advocates succeeded in destroying the institution that had designed, developed and managed the best communications system in the world. In the process, they destroyed the mechanism that could have given all consumers equal access to the high speed services that we so desperately need as a nation. And - in the sacred name of consumerism - they have increased the cost of local telephone service. The consumer has been screwed. Prices are up. Service quality is down. Network extensibility has evaporated.

Why do we want this?

This is not a new issue. People with vision have known for years that what we are doing is dumb. Our existing networks are overloaded. They are deteriorating. The existing copper infrastructure will never support the bandwidth demands of distance learning, instantaneous telecommuting, Internet multimedia and interactive services. We need fiber to the home.

It is time for us to face this issue head-on. We must embrace the concept of universal broadband access for all consumers. We must bury our differences and get on with the program.

We must rebuild the Bell System. It will take an organization of that size, that focus and that discipline to get the job done. Dedicated management assisted by top down engineering to establish standards, technology initiatives and service criteria. A new network designed to provide safe, secure Internet services. Yes, it must be a regulated public utility. Yes, its business must be limited to the provision of transmission bandwidth and communication services. And yes, its facilities must be open to any company that wants to provide a networked service - including television programming and Internet access. And yes, we need to set goals, objectives and criteria for the Bell System that mesh with what we want to accomplish.

There will be, of course, opposition to rebuilding the Bell System. From individuals who have a selfish-best-interest at stake. From those who hate big companies - in any form. Most of the detractors, however, will be people who do not understand the fundamental economic and structural requirements of universal broadband service. We need to provide them with a comprehensive response.

We need leadership. Not politics. We need vision. Not chaos. If we are to remain a prosperous and competitive nation, if we want to minimize the impact we have on our environment, if we want to create equal access to the vast riches of the information age - then we must structure a vehicle to get us there.

The easiest, least costly, technically sound and financially viable way- rebuild the Bell System.

So environmentalists. Do you really want to reduce the consumption of carbon fuels?

Then reduce the need for private or public transportation.

> Bring the work to the individual, not the individual to the work.

Result: This is another good way to restructure resource consumption in any industrialized nation.[33]

Establish International fuel standards.

Oil companies claim that many refineries, over their life cycle, often do not cover the cost of their capital. There are two main reasons. The first is that in most markets, capacity (in the past) has generally exceeded demand, reducing the price value of refined products and thus

---

33 Written in 2000, the concepts of this article apply to every industrialized nation.

driving down their return on invested capital. Secondly, the refinery business model must balance high fixed costs with low variable costs, frequent product specification changes, and volatile changes in product demand.

For example, there is an increasing focus on the allowable level of sulfur in all fuels. In the primary western markets, Europe and North America, environmental regulation dictates that the sulfur content of gas and diesel fuels must be brought down to 15 - 30 PPM by 2007 and to 10 - 12 PPM by 2012. A number of regions are even moving to 'zero sulfur' fuels with a sulfur content of less than 10 parts per million. We can expect that Asia/Pacific fuel specifications will eventually match those of the western regional markets.

As fuel specifications tighten, however, it is increasingly likely that future fuels can not be made from available oil reserves without substantial modification. On the one hand we have the problem that the world's remaining oil reserves will have increasing sulfur, heavy metal and chemical contaminants. On the other hand is the demand for increasingly "green" fuels. Heavier crude oil feedstock's, with an API of 24 and lower, will become more common. Refiners, caught in the middle, will have to focus more research on the development of synthetic fuels in order to bridge the gap.

Frequent product specification changes and variances in specifications from region to region almost guarantee greater production volatility. We actually waste the energy potential of the oil we refine. If we continue to insist on refining multiple feedstock's, using multiple fuel standards, for distribution to multiple geographic markets, then shortages caused by fuel product manufacturing problems are inevitable.

Why do we want to exacerbate the already tough challenges of an oil depletion crisis?

We can fix this predicament by establishing international fuel standards that balance the need to minimize pollution with the emerging requirement to maximize the energy we get from a barrel of oil. In the past, our focus has been on reducing the pollution caused by burning oil. In the future, our focus must shift to burning less oil. The less oil we burn, the less we pollute the air.

Result: International fuel standards will lead to less volatility in the supply of gasoline and a more efficient use of available oil resources.

Reduce the population.

Here is a challenge for the Sierra Club.

Homo sapiens is an endangered species. If we want to save the environment, then let us focus our collective attention on saving the environment of Homo sapiens.

This challenge is based on one simple fact:
By 2022 we are expected to have more than 8 billion inhabitants
on this planet.
We can not support them all with our dwindling resources of energy.

And we are not prepared for the consequences.

I remember reading - years ago - about an experiment with rats conducted by some University or other. The idea was to see how overcrowding would affect rat social behavior. A small colony of rats was placed in a confined space. Their social behavior was carefully described. Periodically, more rats were added to the box, and the resulting behavioral changes noted. Sure enough, as the population grew, there were significant changes to social behavior. At the extreme, there was a marked increase in aggression, irritation, nervous activity, and compulsive behavior. There was also an increase in the types and severity of disease.

Whenever I think about human population growth, I have to wonder - are we imitating the behavior of those rats I read about so long ago? Our resource problems are not just about oil or natural gas or coal, they include drinkable water, timber, minerals, arable land and marine life. And space.

Should we be concerned?

We humans refuse to introduce effective population policies. For instinctive reasons, population control is an emotional issue. Opposition is supported by cultural attitudes, including the belief that unlimited reproduction is ordained by God (or Allah). Logical arguments based on available food, water and resource deterioration are brushed aside. The impending disastrous effects of overpopulation are thus a given. The oil depletion crisis merely serves to accelerate their arrival.

Take a look at the chart below. It tells a remarkable story if we examine the premise for the data. We know that our energy consumption has roughly paralleled our population growth. Although water power, and then coal, were the dominant forms of energy into the first half of the twentieth century, by the 1950s, oil was fast becoming the fuel stock of choice. Oil consumption increases faster than population from 1950 through 2000, reflecting the growth of energy intensive economies in Western Europe, North America and the Pacific Rim. This Best Case Scenario also projects that overall oil consumption will continue to increase until 2020 as China and India (among other countries) develop more energy intensive economies. By 2020, there will be more than 8 billion people on our planet.

134

But wait. Even in the Best Case scenario, oil production, and hence consumption, is projected to stall around 2020 and then begin to decline. On the other hand, the world's population will continue to increase. This raises a serious question. If the world's economy and its consumption of energy have roughly paralleled the growth of the world's population, but the available energy begins to decline, then what happens to subsequent population growth and the health of our global economy?

The short answer. Something has to give.

Since we humans have developed the remarkable ability to make babies even in the face of horrifying poverty, then we can expect that the decrease in oil production will be accompanied by ever increasing destitution. Worldwide. Overwhelming poverty encourages disease, famine, war and lethal misery. The biggest shock will be in those nations that currently have energy intensive economies. The United States, Canada, Western Europe, Japan, Australia, and so on.

And we must remember. This is the Best Case Scenario. The potential devastation of the Production and Political Crisis Scenarios discussed in this report would cause far greater economic dislocation.

Figure 31
Oil Consumption and World Population
1950 - 2042
Best Case Scenario

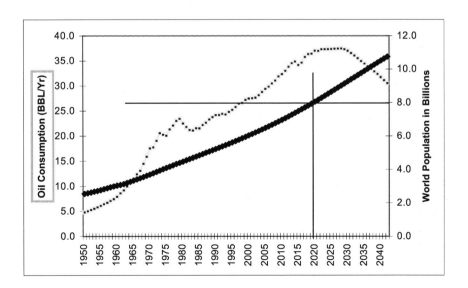

Is there a way out? Sure. Develop alternative energy systems to replace our consumption of oil and natural gas. But the conversion from a petroleum based energy system to an alternative energy system will take from 15 to 20 years. We have to select, research, develop, manufacture and distribute the technology we plan to use. That will take time.

And time is running out.

Even if we are successful in deploying an alternative energy system (no sure thing) population management will still be crucial to avoiding the devastation of declining energy supplies.

In his book, Overshoot, The Ecological Basis for a Revolutionary Change, William Catton claims that the carrying capacity of planet earth has already been exceeded. There is a finite maximum carrying capacity of natural resources - including hydrocarbons, water and arable land. We humans have drawn down on our resources to a point where there will be an overshoot of population - a point where growth exceeds the earth's carrying capacity. This will inevitably trigger a crash in which the population must go through a period of die-off.

In the final analysis, the only sure way to reduce oil consumption is to reduce the population of this planet. The coming economic poverty of an oil starved world simply can not support our planet's existing population of 6.3 billion souls, let alone provide for any population growth. We can either try to manage a "soft landing" or let nature take its course. Doing something means encouraging new attitudes about birthing and the privilege of parenthood on a worldwide basis. If we do nothing, the odds are very good that by 2035 - 2045 unresolved energy shortages will cause sufficient human suffering to force the curtailment of population growth.

It's called "dieoff". The four plagues of the 21st century.

We humans are stuck with obsolete cultural beliefs -
reinforced by our religious institutions.
We promote unlimited birth without regard to the consequences.

That must change. Although the challenges of population reduction will be controversial, they are certainly preferable to massive starvation, disease and death.

Result: The greater the reduction of population, the closer we move humanity to a self-sustaining resource base.

Land reform.

We humans can not expect to conserve energy if we insist on the creation of energy intensive cultures. A key element of conservation, both of energy and the environment, is the allocation of land to support long term conservation goals. Land must be distributed according to its best use in a plan that assumes a reduction of population and a desire to minimize energy demands. To be effective, this type of planning must include Federal, State, regional and municipal law based on carefully scripted guidelines. Zoning, at all four levels of government, is then used to establish the rules of land use. Land is thus set aside for open space, water shed, recreation, timber, food production, residential, retail, office, and industrial development. Included is the creation of transportation corridors to link people with shopping, jobs and recreation. Residential housing and employment opportunities are considered within the context of minimizing their impact on the transportation system.

The long term goal of this effort, of course, is to permit continued prosperity that is consistent with the new realities of energy consumption.

Result: Reducing urban sprawl curtails the growth of energy demand.

Make A Realistic Assessment Of Our Energy Options

Our energy decisions must be made on the best constructive use of each available resource, not on the basis of political correctness or environmental pandering. We have too much at stake. We can not afford to replace thoughtful analysis based on facts with emotional idealism. That means, for example, that coal is a better resource for power generation than natural gas. That also means that if we have 820 GTOE of oil energy left on this planet (as some have suggested) and 15,520 GTOE of nuclear energy available to use, then it makes sense to explore how fission fits into our overall energy strategy.

Environmentalists face a brutal reality.
The halcyon days of clean air are over.

It is only a matter of time. When people get cold, they will scramble for anything that burns. Who wants to watch their loved ones freeze to death? It will not matter what fuel is used to make electricity. Heat and power will take precedence over clean air. Or clean water. Or the three toed red nosed frog. We therefore need to make a methodical examination of our energy options.

Natural Gas.

The depletion of the world's natural gas deposits will eventually mimic the oil depletion scenarios we have constructed. Environmentalists automatically select natural gas as the fuel of choice over coal and nuclear power for the generation of electricity. Gas is preferred for commercial and consumer heating and cooking operations. And the preference for natural gas has spread to Asia, adding to the strain on world supplies.

We just do not have the data to make a reasonable estimate of when our natural gas heritage will run out. Perhaps it will last 35 years. Perhaps 45 years. Although natural gas is an abundant resource, much of it lies in isolated geologic structures, far from the pipelines that could be used to carry it to the user. Liquefaction has thus become a focus for natural gas technology because a portion of our natural gas requirements can be met by transporting it to market as a liquid from remote sites.

> If we waste or misuse our "clean" resources,
> then only the "dirty" ones will be left to meet our energy needs.

In the United States and Canada, we have obviously reached the point of diminishing returns. Doubling the number of drilling rigs over the last 20 years has only produced a marginal increase in natural gas production. New well reserves, on average, have fallen over 80 percent. Should we be using natural gas for electric power generation? Would it not be better to save it for residential heating? If natural gas is not reserved for this use, consumers will be forced to rely on coal, wood, crop residue and animal waste to heat their homes.

The cost in disease and air pollution will be awful.

Coal.

China, the former Soviet Union, the United States, Australia, Germany, Poland and South Africa are among the countries with large coal reserves. Perhaps we have 100 years or more of it left. Coal can be turned into synthetic hydrocarbons - fuel for vehicles, home heating and power generation. Diesel fuel manufactured from coal looks especially appealing because it burns with less pollution than oil based diesel fuel. Coal makes sense for power generation because emerging pollution control technology can be incorporated into the capital cost of newer power plants. So we can not afford to turn up our noses at the thought of coal as an energy solution.

Oil Shales and Sands.

Oil is trapped as a sticky goo in sands and shales. The world's largest deposits appear to be in Canada and Venezuela. World resources may exceed 1.6 trillion barrels of recoverable oil. But recovery operations use oil and lots of heat. In Canada, for example, the producers are using locally available natural gas to heat the sand or shale mixture until the oil turns to a fluid that will seep out of the mix. But is this the best use of natural gas? Nuclear plants can be located in remote areas that are close to these deposits. Could steam generated by nuclear power be used to accelerate the production of oil from oil sands and shales? Can the oil that resides in these deposits be liberated without the environmental damage of excavation? Considering the required investment, technology, and net energy gain, can we develop a realistic estimate of probable production?

Methane Hydrates.

Some believe that methane hydrates on the ocean floor could be a huge source of energy. Hydrates store immense amounts of methane, but how they fit into the energy picture is very poorly understood. Gas hydrates occur abundantly in nature, both in Arctic regions and in marine sediments. Gas hydrate is a crystalline solid consisting of gas molecules, usually methane, each surrounded by a cage of water molecules. Methane hydrate is stable in ocean floor sediments at water depths greater than 300 meters, and where it occurs, it is known to cement loose sediments in a surface layer several hundred meters thick. Extraction of methane from hydrates could provide an enormous energy and petroleum feedstock resource. In addition, there may be conventional gas resources trapped beneath methane hydrate layers in ocean sediments.

Unfortunately, we have not developed the technology needed to "mine" methane hydrates, nor do we really know if these resources are a practical addition to our energy needs.

Hydrogen.

In a very real sense, hydrogen is a manufactured product. Oil, coal, natural gas or nuclear fuel must be burned to produce the electrical energy that is used in the production of hydrogen. At each step of the transformation, one loses about 20% of the initial energy. The indirect or direct conversion of a hydrocarbon fuel into hydrogen does not yield a net increase in energy.

The comparative efficiency of converting crude oil to gasoline is 82 percent, versus 86 percent if crude oil is converted to diesel fuel. Converting natural gas to hydrogen, including the cost of CO_2 sequestration, has an efficiency of 55 percent, while deriving hydrogen from electrolysis has an efficiency of 30 percent. At a production cost of eight cents per kWh (which is consistent with fossil, nuclear, wind and biomass generation), the derivation of hydrogen from electrolysis is approximately four times more expensive than fueling a vehicle with gasoline. In addition, the price of a vehicle fueled with hydrogen fuel cell technology is currently well beyond the financial reach of the typical consumer.

These facts make Hybrid-Electric vehicles very competitive with hydrogen powered vehicles. HEVs can achieve gasoline fuel economies of 30 to 70 miles per gallon (depending on the weight and horsepower of the vehicle) at a modestly higher cost than conventional combustion engine powertrain vehicles. And although the emission characteristics of hydrogen fueled vehicles would appear to make them environmentally attractive, the "upstream" carbon dioxide emissions generated during the hydrogen manufacturing process make hydrogen - as a fuel - only marginally more effective at reducing system emissions than using gasoline in hybrid vehicles.

Obviously, the hydrogen powered vehicle is not a near term solution to the oil crisis. The fuel is expensive, difficult to handle and has a low energy content. (See the discussion in Appendix 1). Hydrogen made from petroleum is a waste of energy efficiency and does nothing to solve the depletion problems of an oil crisis. It is questionable if we could build enough nuclear plants, wind farms and solar panels to enable the use of hydrogen as a replacement for oil.

I quote from the Executive Summary, A National Vision of America's Transition to a Hydrogen Economy - To 2030 and Beyond, by the United States Department of Energy, February 2002.

"The "technology readiness" of hydrogen energy systems need to be accelerated, particularly in addressing the lack of efficient, affordable production processes; lightweight, small volume, and affordable storage devices; and cost-competitive fuel cells."

In other words, we currently do not have a way to manufacture, store, transport, distribute or use hydrogen.

The hydrogen option needs serious work.

Nuclear power.

Environmentalist objections not-with-standing, nuclear power will play a role in man's future use of the earth's scarce energy resources. Nuclear power comes from the fission of uranium, plutonium or thorium or the fusion of hydrogen into helium. Existing power plants typically use uranium. The basic fact is that the fission of an atom of uranium produces 10 million times the energy produced by the combustion of an atom of carbon from coal. There are over 375 commercial nuclear reactors in the world. There is ample uranium. In addition, spent fuel rods contain plutonium and uranium which can be separated in a reprocessing plant and used as reactor fuel. We can build breeder reactors that will actually add to our nuclear resources. Despite activist hysteria, the environmental penalty is manageable.

But we need a long term plan that uses this resource wisely. We must take our time, assemble the best technology we can develop, and deploy it with care. My fear is that environmentalists will prevent the intelligent use of nuclear power. And they will be successful until there is a worldwide energy crisis. In the ensuing panic, politicians will scramble to approve the construction of nuclear plants without adequate engineering or management. Environmentalist objections will be swept aside. The result will be an invitation to nuclear disaster.

Is that what we want?

If environmentalists are concerned about the use of nuclear power, they will better serve their cause if they insist on the prudent use of nuclear power. Starting now.

Clean coal technology.

Fischer-Tropsch Gas-to-Liquids (GTL) technology has shown promise in public transportation, personal vehicle, and fixed site engine applications. GTL diesel is not just a cleaner version of refinery diesel; it is a water-white, gas-based fuel, that is virtually free of sulfur and aromatics content. This gas-based fuel can be used in existing diesel engines and made available through existing channels of distribution. GTL ultra-clean diesel can significantly reduce the emissions of particulates and carbon monoxides. Engines running on GTL diesel fuel create less smog. GTL fuels can be used in standard diesel engines, either as a blend with standard diesel fuel or as a pure fuel for advanced diesel engine designs or hydrocarbon-powered fuel cells.

Result: We need to determine if clean coal technology can provide us with a viable alternative to conventional diesel fuel for mobile transportation and fixed site applications. If this works, we could move a substantial percentage of vehicles (see Figure 31 below) and power plants to a coal based diesel fuel.

Solar power.

We have only begun to understand the resource potential of the solar energy option. The cost of electricity produced by photovoltaic

(PV) technology has fallen by 90 percent since the early 1970s.[34] Photovoltaics are producing electricity for critical loads from the polar ice caps to the tropics. There is a strong market in developing countries to provide rural electrification with solar panels, replacing kerosene lamps at a far lower cost than central station power plants. Solar power has already proven its value in applications where the power consuming device must be located far from the electrical grid.

Solar power technology, however, is still in the stone age. Hyperbole aside, we can do better.

Photovoltaic cells convert sunlight directly into electricity. To manufacture a photovoltaic cell, a material such as silicon is doped with atoms from an element with one more or less electrons than occurs in its matching substrate (e.g., silicon). A thin layer of each material is joined to form a junction. When photons from sunlight strike a PV cell, these mismatched electrons are dislodged. The electron creates a current as it moves across the junction. The electrical power thus generated is gathered through a grid of physical connections. Various currents and voltages can be supplied by manufacturing an array of cells.

The DC current produced depends on the material involved and the intensity of the solar radiation incident on the cell. The most widely used material is the single crystal silicon cell. The source silicon is highly purified and sliced into wafers from single-crystal ingots or is grown as thin crystalline sheets or ribbons. Polycrystalline cells are an alternative. Although inherently less efficient than single crystal solar cells, they are cheaper to produce. Gallium arsenide cells are among the most efficient solar cells, but are expensive. Another approach to producing solar cells that shows great promise are thin films made from amorphous silicon, copper indium diselenide or cadmium telluride. Thin-film solar cells require very little material and can be easily manufactured on a large scale. Manufacturing lends itself to automation and the fabricated cells can be incorporated into building components.

But we must expand our search for solar power solutions. For example, STMicroelectronics, Europe's largest semiconductor maker, has indicated that by the end of 2004, it expects to be able to manufacture the first stable prototypes of a new solar cell made from organic materials. The French-Italian company expects cheaper organic materials such as plastics to bring down the price of producing energy. Over a typical 20-year life span of a solar cell, the cost per watt could be as little as $0.20, compared with the current $4. The new solar cells would even be able to compete with electricity generated by burning fossil fuels such as coal, oil and gas, which costs from $0.20 to $0.80 per watt.

---

34 Information courtesy of the Solar Electric Power Association (SEPA).

Global conditions, including cloud cover, topography, latitude, and altitude, mean that solar power will always be a variable resource. However, the National Renewable Energy Laboratory (NREL) believes there are suitable sites for solar power generation all over the globe.

Barriers to widespread use of solar power include the development of suitable energy storage devices (including batteries and hydrogen converters), panel efficiency, and manufacturing cost. Of particular interest is a system that permits solar power to generate electricity that is then used to manufacture hydrogen. At night, a fuel cell converts the hydrogen into electric power.

Although we humans are a long, long way from being able to use solar power as a replacement for oil or natural gas fired electricity generation, it is an option worth more effort.

Move Aggressively To Utilize New Fuel Technologies

Strategy

We must focus our attention on the development of alternative energy systems. The use of the word "systems" is important. All energy comes from nature. It must be found, produced, refined, transported, and delivered. There must be a mechanism for consumption that can use the energy we produce.

Thus each alternative source of energy comes with a set of metrics. How will it be produced? Transported? Distributed? Consumed? And if there are environmental impacts, then how can these be mitigated?

Most of the alternatives assume the creation of a mass transportation oriented economic structure. Energy will be delivered in the form of electricity because the energy source is not mobile. Individual mobility will thus be restricted to available transit vehicles - trains, subways, busses and light rail systems because these vehicles can utilize electric motors. There must be a gradual migration of personal and commercial vehicles to hybrid and GTL diesel technology.

In the following chart, we have calculated the achievable conversion of automotive fuel technology to hybrid gas/electric, diesel, and hydrogen powered vehicles. If this were done on a worldwide basis, oil consumption could be reduced by up to 33 percent.

Figure 32
Fuel System Evolution

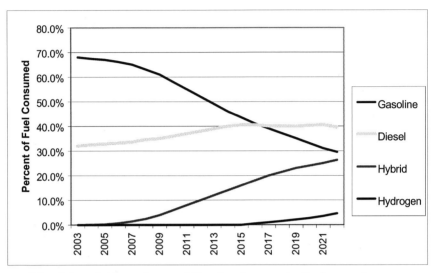

Oil crisis solutions demand the development of advanced technology. We must evaluate our options and then focus our attention on developing the systems necessary to deliver energy to the commercial and individual consumer.

Hopefully, an open and frank discussion of our energy needs will lead to the development of an energy policy as well as the means of achieving our goals.

Develop a new fuel system.

The exploration, discovery, production, refining, transportation, distribution, and consumption of oil is a "system" of interrelated parts. Crude oil becomes gasoline, heating oil and feedstock. Vehicles and furnaces are designed to burn oil. There would be no sense in refining oil if there were no way that a consumer could acquire and use the resulting product.

We must first replace oil as a vehicle fuel because this is where 70 percent of the oil is used. The replacement of oil as a vehicle fuel requires the development of a new fuel system. Included in this effort will be the identification and production of raw materials, the manufacture of the fuel, solutions to the challenges of storage and distribution, and a means of consumption. It is a conversion process - from raw materials to energy.

The development of a reliable, available, inexpensive and safe fuel system will require research, engineering, production and marketing

144

expertise. Big bucks. Lots of coordination among thousands of people. Corporate infrastructure. Research that evaluates the inevitable pros and cons of several alternative fuel systems. And most of all - a sense of mission.[35]

Will the solution be hydrogen in some form? Innovative C technology[36]? Solar power? Clean coal? Nuclear power? In truth, the solution probably lies in bringing together a number of these components.

How do we do this?

By gathering together the best minds we can find on this planet and focusing their attention on a single objective. The DOE has already made an excellent start by funding multiple energy research programs. And Congress has been willing to throw money at various energy initiatives - largely because it was the politically correct thing to do. But these efforts present us with a timing problem. They are too little and - probably - to late.

Which brings us to NASA.

Sorry. But I can not get excited about another trip to the moon or yet another pretty picture of mars. Not when we humans should be focusing our attention on urgently essential solutions to real problems here on earth. NASA was created to find a solution to a specific challenge - go to the moon (and elsewhere). But that mission is over. So now we have an institution that no longer serves an essential purpose. NASA desperately needs a new assignment.

Why are we spending tax dollars on space
when we need heat and power here on earth?

So here is a proposal. Let's remission NASA. Change the name. Meld its resources with existing DOE programs. Focus its entire attention on the development of a new fuel system. Bring in the best minds from all over the world. Make this an international effort.

Now NASA would have a really, really useful purpose.

Make sense?

---

35 In the United States, the Department of Energy has already initiated some very interesting programs.
36 C technology refers to modification and optimization of the hydrogen carbon chain to produce a molecule that contains easily released energy.

Oil companies are part of the solution.

I am not a fan of American President George Bush 2. One of the reasons is that he and Vice President Cheney are much too close to the oil industry.

But for the purposes of this report my rationale doesn't have anything to do with insider influence. It is based on a recognition that oil companies are bureaucracies - huge, insular, inflexible, self-serving, highly politicized, bureaucracies. With the possible exception of BP, there is little evidence of the leadership and vision it takes to institute creative change. It is also apparent that neither George Bush 2 nor the oil companies have a clue when it comes to marketing. They know how to sell petroleum products. But that's about it.

So they don't know how to plan for the future, they don't know how to establish a credible energy strategy, and they do not know how to communicate where we are going. That's too bad. The world desperately needs an energy system strategy. A credible energy system strategy. With money and resources to back it up.

Big oil - if it gets its act together - should become a key component of an international energy strategy. Like it or not, this is a huge, expensive, program and that means a huge and complex organization will be required to get the job done. A new energy system will place incredible demands on research, development, manufacturing, and distribution. It will need a very big - and experienced - manufacturing infrastructure. Eventual deployment will be easier if it is made through an already established channel of distribution. Someone has to deal with comprehensive dealer training and support, supply depots and transportation, as well as manufacturing facility construction, maintenance and management. That takes big bucks and a working organization.

It also needs big numbers of very dedicated people with a clear vision of their task. A reorganized bureaucracy with advancement and bonuses tied to program milestones. Pride. Accomplishment. A sense of destiny. Oil companies are populated with some very bright and creative people. Free them from the bondage of existing organizational structures. Challenge them to work for our future.

Government policy can help by ensuring that it is in the oil companies selfish-best-interest to pursue the development and deployment of a new energy system. That means dumping the existing pork filled bills in Congress. Pork only extends the life of obsolete business strategies. Oil needs to establish and pursue a new direction.

And it must be an international program. Financing and brains from multiple countries. Funded by adding a special tax of a few pennies on each gallon or liter of gasoline sold at the pump. All pay as you go. No more bureaucratic, politically expedient pork.

This has to be organized as a partnership with a remissioned NASA and a group of international participants. And then - tell people what this organization is doing.

Can Big Oil become a player?

Maybe. But not without a good swift kick.

Peace Through Communication

The single largest obstacle to engineering a "soft landing" for the oil depletion crisis is the enormous energy of Islamist fundamentalism. As discussed in Chapter 5, this is a cultural problem that dates back to the Arab colonial period, 1918 - 1939. The roots of Jihad predate the 7th century.

As the civilization of the Roman empire slipped backward into the horror of the Dark Ages, Arabia emerged to become a dominant political and cultural force in the Mediterranean. The Saracenic empire was forged by conquest from 630 to 750 AD. At its zenith, it extended along the southern Mediterranean Sea from Spain to the borders of India. Among Arabs, Muslim religious influence was a key contributor to the adoption of a common language and a reasonably cohesive set of cultural beliefs. It was an era of great progress in the arts and sciences. It was also an era of remarkable tolerance and it was this tolerance that encouraged intellectual achievement to flower within the Saracen empire. Many of Islam's Caliphs became enlightened patrons of learning and it can be said that Saracen cultural progress exceeded that of Christian Europe until well into the 11th century. Cultural conflict between the peoples of the Middle East and Western Europe again erupted with violence during the Crusades.

Unfortunately, the Saracenic empire declined very rapidly. The Arabs lacked the political experience to forge a cohesive federation. Strife between Sunnite and Shiite factions weakened the influence of Islam, the core of their political system. In 750, a Shiite revolt succeeded in moving the throne to Baghdad. In 929, members of the Ommiad family succeeded in establishing an independent caliphate at Cordova in Spain. Descendents of Mohammed soon proclaimed themselves the independent rulers of Morocco and Egypt. By 1057, the caliphs in Baghdad had surrendered all of their temporal power to the Sultan of the Seljuk Turks.

The political systems of the Northern Mediterranean progressed from tribes and clans to feudal rule, then coalesced into regional empires that eventually became modern nation states. Despotic rule was replaced by egalitarian political concepts. Cultural values evolved from feudal, monarchy centric mores into democratic concepts of equality, justice and the rule of law. Within the industrialized and emerging nations of the world, this is still an ongoing process.

Unfortunately, Islamist cultural tradition has not gone through the same transition. We can understand this fact if we examine the basic themes that emerge from extremist doctrine:

- Islam is the State and will provide political leadership for all Muslims;
- intellectual discussion must center around approved Islamic thought;
- there is no merit in adopting western values or technology;
- only Muslims are true believers in Allah;
- Jihad will lead to victory over the forces of evil;
- and all non-Muslims are infidels (unbelievers, atheists, heathens), whom Allah will punish because unbelief is the greatest sin of all.

Fundamentalist beliefs encourage exclusivity. An us versus them mentality. Parochial conviction prohibits tolerance or the pursuit of knowledge. Closed social structures foster a deep suspicion of outsiders who are treated with distrust and hostility.

Thus there is a clash of wills and ideas. On the one hand we have the nations of the western hemisphere with a fragmented Christianity, a declining commitment to the ideals of democracy, and deteriorating moral values. On the other hand we have an increasingly vigorous Islamic minority who are very secure in their beliefs, hate the idea of liberal democracy and are ready to enforce their concepts of morality on everyone else.

So we have the war on terrorism. And it will do some good against those who believe that murder is the route to salvation. But if we want to bring about a lasting and beneficial peace in the Middle East, then we must develop a way to constructively deal with this surge of Islamist volatility while we build a bridge of understanding to the Arab world. The majority of Muslims, like the majority of Christians, understand that economic and social progress is the result of constructive endeavor. Peace, compassion and friendship is the preferred framework for social discourse. Although our philosophical outlook differs, we share many fundamental beliefs about self , family and community.

Unfortunately, a vocal minority apparently believe that destruction must precede progress. Ignorance is pervasive. Deception is expected. The industrialized world doesn't help itself by treating Arabs as though they are second class citizens. In Saudi Arabia, for example, the United States appears determined to make the same political mistake it made in Iran. We curry favor with the rich and powerful while we ignore the common man. That policy will backfire and explains why fundamentalist zealots hold the key to Saudi Arabia's future.

And the flow of oil.

We are in a trap. All of the proposed mobile energy systems will take from 15 to 20 years to put in place. We need to buy time. We must maximize the exploitation of our existing resources while we develop and deploy new energy solutions. That requires political stability in the Middle East. Pakistan, Iran, Iraq, Egypt, and Saudi Arabia are the key states to watch. If they all become Islamist theocracies, then we must expect a period of world-wide economic deprivation. Perhaps world war.

Centuries of adversarial distrust will be difficult to resolve. Armed conflict will not bring about lasting peace. It is unlikely that the dogmatic ideology of opposing religions will be helpful. Politically motivated government diplomacy is unlikely to bring about mutual understanding. Officious bureaucratic bungling works against us. Cultural change takes time.

And time is running out.

So what can we do?.

We must encourage the development of personal friendships.
Hatred seldom lasts between people who have laughed together.

The key word is communication. The people to people kind. Christian to Muslim. And Hindu. And Buddhist. And anyone else who wants to join the quest for peace. Outreach based on love, tolerance, friendship and fellowship. A passion for truth that ascends above ideology. Patience and persistence.

Western institutions of higher learning - colleges and universities - need to implement programs that help Middle Eastern students to make a relatively painless cultural transition to the unstructured environment of western campuses. Academia and the student population must participate in the effort to build a bridge between Muslim and western philosophy. Peace though communication can start on the campus.

Ultimately, it will be up to a new generation of young people
to persevere in the quest for mutual understanding.

It's going to be your world. What do you plan to do with it?

Peace Through Communication sounds far more attractive than the alternative.

If we treat each with respect, will understanding follow?

Chapter 10 THE QUESTION

It's up to you.
Your destiny hangs in the balance.

Denial

Politicians will deny there is a problem. Journalists will reject the hypotheses. Idealists - be they liberal or conservative - will greet the conclusions of this report with cynical comment. This is - after all - bad news. The ramifications of an oil depletion crisis will change the world in ways we do not like.

This report will be criticized. The assumptions, calculations and data will be challenged. Cynics will cling to ignorant beliefs. It doesn't matter. Unless we can find another 1 Trillion barrels of oil over the next 20 years, an energy crisis will occur before 2022.

Probably much sooner. We appear to have maximized our ability to produce conventional crude oil. Alternative oil resources will be difficult and expensive to exploit.

We humans are in this together. That's us. People. Kids and adults. Family and friends. Every nation, every race, and every ethnic group. The potential progress of civilization hangs in the balance.

Pointing the finger of retribution will not be helpful. Ideological confrontation will not be constructive. The blame for our predicament rests with multiple peoples in multiple nations. If we want to mitigate the impact of the impending crisis, we must work together. - Conservative and liberal, believer and agnostic, rich or poor, young and old - in every nation.

Can we humans come together? Can we avoid the dreadful impact of oil depletion?

I present two possible scenarios [37] .

The Peace Scenario

Before the end of 2005, a group of nations join together in an effort to ease the transition from an oil based world economy to one that embraces an alternative form of energy. It is a cooperative effort. An international consortium addresses the research, development, manufacturing, distribution, storage, handling, transportation and consumption issues of a new energy system. Solutions are found and implemented. By 2015, the new system is ready to deploy on a world-wide basis.

---

37 There are obviously multiple scenarios. But they will all replicate some variation of the two scenarios described here.

In the meantime, consumer and producer nations have worked out an oil supply agreement. It includes a resource depletion timetable that will guarantee declining, but not disruptive, levels of production. Consumer nations agree to fund additional drilling and recovery programs. Producer nations address long term maintenance and environmental issues.

It is recognized that the energy crisis is also a population crisis, and efforts are made to reform human attitudes toward birthing. Most of the recommendations described in this report are carefully evaluated. Many are implemented in some form. Measures are taken to defuse Islamist extremist activity. All over the world, colleges and universities implement programs to bridge the cultural gap between the Middle East and a global population. Although the process is painfully slow, some form of democracy - or managed democracy - spreads throughout the region.

The energy crisis is brought to a peaceful conclusion. There is a relatively mild recession during the transition from a petroleum based world economy to one founded on a new energy resource.

The Conflict Scenario

Belligerent behavior, uncompromising ideology and plain ignorance paralyze the management process of contemporary political institutions. Politicians are consumed by an increasingly acerbic confrontation. Liberal versus conservative. Us versus them. Cynical commentary parades as moral righteousness. The maintenance of political power remains more important than legislative wisdom.

There is little cooperation between consumer and producer nations. Governments continue to operate on the principle that the preservation of political power is far more important than making rational decisions based on objective research and erudite analysis. The satisfaction of selfish-best-interest takes precedence over rational inquiry. Ideology is more important than truth. Few politicians have the vision or will to implement the recommendations found in this report. Increasingly dysfunctional national governments are incapable of implementing the programs we so desperately need in order to survive the economic holocaust of an oil depletion crisis.

All attempts at creating a global oil production agreement are met with charges of imperialism. Oil exploration accelerates to a frenetic pace. Military force is used to ensure territorial domination. Terrorist activity disrupts oil production and plunges the world into economic chaos. Only sporadic and fragmented efforts are made to develop a new energy system. No single energy solution emerges. Efforts to mitigate the population crisis are met with uncompromising ideological criticism. Bridging the gap between Middle Eastern and global cultural values is dismissed as racist. Theocratic or secular dictatorships continue rule throughout the Middle East.

By 2020, worldwide economic activity and cultural progress have been sharply curtailed by the lack of energy resources. Endemic recession and the four plagues of the 21st century have become a painful reminder that we have failed to deal with oil and Jihad. Millions die.

The End

So I end this report with a question.

Which of these scenarios is a more likely description of human destiny?

Appendix 1. Supplemental Information

Congressional Research Service

The following report, completed in 1995, is one of the most comprehensive and candid reports available on the subject of oil depletion. By inference it shows that America's Congress knew, or should have known, about oil depletion by late 1995. Yet these politicians have done virtually nothing to address the problem.

Why not?

World Oil Production After Year 2000: Business As Usual or Crises?

Joseph P. Riva, Jr.
Specialist in Earth Sciences
Science Policy Research Division
Redistributed as a Service of the National Library for the Environment: 16 Pages

Extract
August 18, 1995

"During the Arab oil embargo of early 1970s, the United States gross national product declined and unemployment doubled. .... Oil (supports) ... world commerce and accounts for 40 percent of total primary energy demand. ... There have been projections by the International Energy Agency (IEA) and the Energy Information Administration (EIA/DOE) that world oil production will have to be increased by about one-third in the next 15 years to meet rapidly rising demand. This would require a major expansion of oil production capacity in the Persian Gulf OPEC countries that is far above their present plans. While the Gulf OPEC countries may have sufficient oil reserves to support ... additional production ... , an expansion of such magnitude would require capital they may not have, and the assistance of foreign financial and oil interests, they may not want.

Outside the United States, where proved reserves are defined primarily on petroleum engineering criteria, there are many vested interests that define oil reserves to meet political or economic objectives.

> Discounting the reserves that may be exaggerated, and utilizing only that portion of the resources that may be produced in actual practice, could reduce the ultimately recoverable oil remaining in the world to a level where the midpoint of world oil depletion would occur at the turn of the century (2000), followed by a production decline of nearly three percent per year [38].

Such a resource driven world oil shortfall is ... not ... amenable to the usual political, economic, or military solutions.

... Significant increases in world oil demand will have to be met primarily from Persian Gulf supplies. This is a region with a history of wars, illegal occupations, coups, revolutions, sabotage, terrorism, and oil embargoes. To these possibilities may be added growing Islamist movements with various antipathies to the West. If oil production were constrained, oil prices could rise abruptly along with adverse world economic repercussions. If the IEA and EIA are correct on the demand side, deficient world oil productive capacity could cause an oil crisis within 15 years .... However, if the increases in world oil demand were more moderate, and there is long-term relative peace in the Middle East, with increasing foreign participation in upstream oil activities, a business as usual world oil demand and supply situation would be a likely scenario ... .

... The International Energy Agency (IEA) projects that the share of oil in the world's energy mix will remain relatively stable at around the 40 percent level to the year 2010. However, since world energy demand is expected to increase, IEA projects that oil demand will rise from the current 68 million b/d (in 1995) to around 76 million b/d in year 2000[39] and 94 million b/d in 2010. ... .

.. The OECD countries are projected by IEA generally to experience declining oil production. The rest of the non-OPEC world, notably China and the republics of the former Soviet Union and Latin America, are expected to increase their oil production. However, even IEA's most optimistic projection of non-OECD, non-OPEC oil production increases accounts for only 20 percent of the estimated rise in total world oil output. Then, 80 percent of the projected increase in world oil production, some 21 million b/d, would have to come from OPEC. Over the next 15 years, the OPEC countries would be expected to increase

---

38 On a rate of change basis, my scenarios appear to confirm that conventional oil production did peak in 2000.

39 Adjusting for the impact of the recession, this estimate has proven to be reasonably accurate.

their oil production from about 27 million b/d to about 48 million b/d or by 78 percent. ...

Considering reserve/production ratios, reserves, and resources, it is evident that the major sources of any increase in world oil output will have to be Saudi Arabia, Iran, Iraq, Kuwait, and UAE (United Arab Emirates including Abu Dhabi, Dubai, Ras al Khaimah, and Sharjah). ...

Little oil development was possible in the Persian Gulf region during World War II, although large fields had been located in Iran, Iraq, Kuwait, and Saudi Arabia. By the end of the war, it had become evident that the Gulf would become a major oil exporting region when adequate outlets became available. In the postwar years, a rapid rise in world oil demand was coupled with a rapid production expansion in the Gulf. With the nationalization of Iranian oil in the early 1950s, Kuwait became the Gulf's leading oil producer, holding this position until 1965. Saudi Arabia's rise to prominence as an oil-rich state came later. It has since, however, achieved preeminence as the largest holder of oil in the world. ...

... Saudi Arabia had granted a concession to the Arabian American Oil Company (Aramco) in 1933. The first discovery was in 1935, ... Initial production was modest. The discovery that transformed the prospects for the oil industry was that of Ghawar in 1948. It is the largest oil field in the world (82 billion barrels), ... In addition to Ghawar, Saudi Arabia was found to contain ten other super-giant fields, including the world's largest offshore field. A national oil company was established in Saudi Arabia in 1956 to conduct the exploitation of petroleum resources outside of the Aramco concession. Since that time, oil production has become increasingly governed by the state. In 1974, the Saudi government purchased a majority participation in Aramco and the company became fully nationalized as Saudi Aramco in 1988.

The Arabian-Iranian basin is underlain by a deep basement of Precambrian rocks ... Overlying the Precambrian is a wide platform area whose strata formed from slow, intermittent subsidence and deposition along sea margins and into shallow seas during Paleozoic and Mesozoic time. The platform consists of a thick sequence of continental and shallow marine sediments that dip gently east and northeast. Tectonic movements, beginning in Late Cretaceous time, finally eliminated the ancient seaway and culminated, at the end of the Tertiary, with the folding of the Zagros, Taurus, and Oman mountains. A number of large regional, north-south anticline trends, that are probably related to basement uplifts, occur in the platform. Many super-giant oil fields (including Ghawar) are contained within structural closures along these trends. To the north and east, the sediments in the deeper parts of the basin were folded along northwest to southeast trends. In northeast Iran, a series of overthrust faults occurs, and the mountain ranges are tightly folded and faulted. Thus, the oil fields of Iran and Iraq are elongated in a northwest-southeast orientation, in contrast to the north-south oriented fields of Kuwait and Saudi Arabia.

The oil fields in the Arabian-Iranian basin vary in their structural and stratigraphic characteristics according to their location. Jurassic reservoirs occur in the broad, gently folded structures of Saudi Arabia. Most of the oil is produced from the Late Jurassic Arab formation which is composed of permeable carbonates alternating with evaporites and sealed with a thick anhydrite. It also contains organically rich carbonates that accumulated under anaerobic conditions as the sea transgressed the region in mid-Jurassic time. These deposits are the likely source for most of the petroleum in the Jurassic reservoirs of Saudi Arabia. Jurassic sedimentation concluded with four main depositional cycles, each consisting of a shallowing marine carbonate sequence overlain by anhydrite. .... The largest oil accumulations occur in the oldest cycle, ... At Ghawar, the north-south trending oil pool is approximately 140 miles long covers 875 square miles. The oil column reaches a maximum of 1,300 feet.

... The extremely large oil accumulations in Saudi Arabia can be attributed to the development of thermally mature, organic rich sediments underlying or adjacent to widespread, highly porous and permeable carrier and reservoir beds, and the cyclic deposition of evaporite seals. The tectonic activity in the region was sufficient to create large structural traps, but not intense enough to disrupt oil migration paths or evaporite caps. The folding is uniquely simple and uniform .... Thus, virtually all of the migrating oil was trapped in the large primary structures, creating huge oil fields but practically no small ones. Almost all other world oil provinces exhibit a much more complex structural character ...

Reported Proven Oil Reserves, Major Middle East OPEC Producers, late 1980s (Bbl)				
	1986	1987	1988	1989
Saudi Arabia	169.2	169.6	172.6	257.6
Kuwait	94.5	94.5	94.5	97.1
Iraq	47.1	100.0	100.0	100.0
Iran	48.8	92.8	92.8	92.9
UAE	33.1	98.1	98.1	98.1
Total	392.7	555.0	558.0	645.7

During the four-year period, the reported oil reserves of four countries increased by 64 percent, or 253 billion barrels. The largest increase was in the UAE (196 percent), followed by Iraq (112 percent), Iran (90 percent), and Saudi Arabia (52 percent). Since OPEC

156

production quotas and levels are partly determined by reserve size, it was more than a coincidence that each country chose this time ... to increase reported reserves. A more important consideration, however, is whether the reserve increases are political or real. ... During the late 1980s, not enough exploration wells were drilled in Saudi Arabia, Iraq, Iran, or the UAE to increase proved reserves by the amounts reported through new field discoveries, nor were large new discoveries made. ...

(In order) to achieve higher recovery rates (50 to 60 percent of the proven reserves), huge capital investments in water injection wells, field wells, pumps, pipelines, and water separation plants would be required. Eventually, the ... fluid would approach 90 percent salt water that would have to be separated and disposed of..... Under the definition used in the United States, the current proved reserves of Saudi Arabia probably would be about 160 billion barrels, leaving some 101.2 billion barrels as inferred reserves that, at very considerable expense, could be .... eventually recovered.

Similarly, Kuwait currently has about 86 billion barrels of proved reserves, with field growth potential of perhaps 10.5 billion barrels, at significant additional expense. In the UAE, about 61 billion barrels may be proved, leaving about 37 billion barrels of inferred reserves potentially available at substantially higher cost. In Iran, perhaps 69 billion barrels of the 89 billion barrels reported as reserves qualify as proved reserves, while in Iraq it is possible that as much as 91 billion barrels of the reported 100 billion barrels of reserve may qualify. Kuwait, UAE, Iran, and Iraq have reported generally level reserves since 1987....

Persian Gulf producers could decide that it would be to their economic advantage if the enormous investments necessary to further increase oil production were not made. Then, as increasing demand forced up world oil prices, they would realize higher income from level production without more rapidly depleting their oil reserves. ...

If World oil demand increases significantly from the current (1995) 68 million b/d, to near the 94 million b/d in 2010 projected by IEA, OPEC would be expected to increase production from the present 27 million b/d to as much as 48 million b/d. (By) ...2010, OPEC would .. control world oil prices. OPEC has no current plans to increase oil production (to these levels)[40]. ... The costs to further increase capacity, ... to the levels projected, would be so high that the international oil industry would have to be involved. ...

The effect on the rest of the world, however, would be much less salubrious. Oil prices significantly affect world commerce. For the industrial countries, the oil price spike of the early 1970s brought profound dislocations and a deep recession. In the United States, gross national product fell by six percent between 1973 and 1975, while unemployment doubled to nine percent. In Japan, gross national product declined in 1974 for the first time since the end of World War II. The economic impact in Europe was correspondingly severe. The oil price increase was an inflationary force that continued when economic growth resumed in 1976. ... developing countries ... reacted by borrowing and going into debt. Thus, their ability to grow economically was retarded, and, in some cases, halted altogether. ...

In any scenario of increasing world oil production, Saudi Arabia must play a leading role. ... an increase in world oil demand, on a scale envisioned by IEA, would require at least an additional 3 million b/d of oil output from Saudi Arabia. Saudi Arabia's ability to fund an additional oil production increase of this magnitude is questionable. The country now runs a yearly deficit and its monetary reserves, once $200 billion, are down to $30 billion. The faltering economy has resulted in a somewhat reduced standard of living, record unemployment, an increasing homeless population, and some civil unrest. Since the Gulf War, the Saudi monarchy has been confronted with a growing Islamist movement that seeks to achieve a comprehensive transformation in the Kingdom's social, economic, and political life. .... fundamentalists goals include an even more strict adherence to Islamic law and practices with the censorship of all foreign (infidel) influences. In addition, they would establish a strong army for protection, independent of western assistance, and would control government spending.

... A worst case scenario would be a long and destructive civil war between the fundamentalists and the House of Sa'ud, fought above the world's largest oil fields. ...

In the United States, the oil industry is unique because onshore mineral rights are mostly in private ownership, and the Security and Exchange Commission enforces rules that define proved reserves in terms of actual development drilling. Then, as fields are extended by additional drilling, reserves are increased. Thus, onshore U.S. fields usually appear to grow over time. In the offshore United States and in foreign countries, a discovery is appraised on the total volume of oil

---

40 As of May 2004, OPEC still has no plans to increase its production to the levels the world would need to sustain projected increases in consumption. Why not?

most likely to be recovered as a basis for development planning. While the first estimate is not always correct, it is as likely to go down as up. There are many vested interests that define foreign oil reserves to meet political or economic objectives. .... In the former Soviet Union, oil reserves often were exaggerated by the inclusion of in-place oil that was neither economically nor technically recoverable.

Discounting the reserves that may be exaggerated and worldwide field growth projected on the U.S. onshore model, and utilizing only that portion of estimated undiscovered oil that may be considered to have a chance of being produced in actual practice, *could* reduce the total amount of ultimately recoverable (conventional) world oil from 2,330 billion barrels to only 1,750 billion barrels. An analysis using production curves from depletion models, with a 1,750 billion barrel ultimate world oil recovery, indicates the midpoint of world oil depletion would likely be reached by about the year 2000 at the about the current rate of production and then decline at a rate of about 2.7 percent per year. If successful efforts are made to delay the onset of decline, the subsequent decline rate would be greater. Heavy oil, tar sand, and enhanced oil recovery will become important after year 2000, and would mitigate but not reverse the world oil production decline. ...

If the IEA and EIA are correct on the demand side, deficient world oil production capacity may cause an oil crisis in less than 15 years. Political disruptions in Saudi Arabia that constrain oil exports could cause an oil crisis at any time...."

National Council for Science and the Environment
1725 K Street, Suite 212 - Washington, DC 20006
202-530-5810
ncseonline.org

Oil Depletion in the United States and the World

Seppo A. Korpela, Ph.D.
Professor, Mechanical Engineering
The Ohio State University
Columbus, Ohio, 43210

A 15 page working paper for a talk to the Ohio Petroleum Marketers Association at their annual meeting in Columbus, Ohio, May 1, 2002. Thank you Seppo for allowing me to use your excellent work.

Extract

"Introduction

Oil production peaked in the United States in 1970 and the world peak is imminent. ... Half of world oil production comes from 120 large fields, many over 30 years old, and most are already in secondary recovery. The rate at which new oil is found has (decreased), with the result that world oil production is set to begin its decline sometime between the years 2005 and 2010. ...

Oil depletion is not yet in the consciousness of the general public, who only notice the rise in prices for gasoline, heating oil, and natural gas, and then are ready to pronounce a multitude of causes, few of which go the heart of the matter. ...

To understand oil depletion one needs to stand back from the confusion of mixed signals that appear in daily papers and TV, whether proffered by journalists or by oil industry spokesmen. The former are generally ill-informed and the latter, by virtue of their position, prone to give a one-sided account. ...

Everyone seems to agree that sometime in the future oil will be exhausted, but the perception is that the day is still far away. This view arises from the common pronouncement that at present rate of production the discovered reserves will last 40 years. Although this is true, it is highly misleading, for it presents a picture in which production is flat for 40 years and then drops to zero after that.

The likely scenario is that oil production increases until depletion overtakes production and then production drops at some specified yearly rate until it is exhausted. When the world enters this downslope, half of the original oil endowment will still be there to be extracted.

This is cold comfort for the consuming countries who are faced with a struggle to get by with diminished supplies each year. ...

Where is the oil?

Oil originates in the sedimentary basins of the world. Particularly important are those basins that have been on the continental shelves for millions of years. Rivers that emptied into them brought effluents composed of silt and remnants of vegetation. In the nutrient rich surface waters of such basins and with the energy of the sun, a profusion of microbial life thrived during certain particular epochs of the geological past. At the end of their short life-cycle these micro-organisms sank to the bottom and there together with other sediment formed layers of carbon-rich organic matter.

Although the continental margins are the most important basins, other large inland bodies of water have also provided a setting for the same life-cycle. The Caspian Sea is an example ....

When the early sediment layers at the bottom of the basin were covered with later sediment, the deeper layers were deprived of oxygen and the organic matter in them did not decay, as it does in ... a kitchen compost. Anaerobic bacteria, however, go to work and turn the organic matter into the substance kerogen. Kerogen can be thought of as immature oil. ...

The earth's oceanic crust is about 5 miles thick and the continental crust varies in thickness from between 25 and 37 miles. The crust is broken into crustal plates below which is a hot molten mantle. This means that temperature increases with depth at a global average rate of 74 degrees Fahrenheit per mile. At the depth from 6000 ft to 13,000 ft temperature and pressure are right for kerogen in the source rock to be cracked into oil. This zone is called the oil window. At depths greater than 13,000 ft temperatures are so high that oil is cracked into gas .

Convection currents in the mantle cause the crustal plates to drift slowly. As they drift, boundaries of the plates collide and on one side folded mountains are formed. On the other side a plate sinks deeper into the earth. In this way, the rock formed in the sedimentary basins may be brought into the oil window and the kerogen in the rock can be converted to oil. At this depth the overburden bears down on the source rock and creates a high pressure. Thermal cracking of kerogen to oil increases the pressure further and to relieve the pressure, oil will migrate to surrounding strata if a path is open for it. All this takes place slowly over a time spanning millions of years. ... much of oil that was created over time must have reached the earth's surface and disappeared into nature. The earliest oil fields were found where there was visible evidence that such a seepage has taken place.

The strata under a range of folded mountains tend to have a shape similar to the mountain itself. Such a feature can be pictured as a

Oil, Jihad and Destiny   161 domed hallway, with the technical term anticline. If an anticline consists of porous and permeable rock and is covered by an impermeable seal, preconditions for an oil field are in place. Oil fields tend to be found not under the mountains, but in the foothills and valleys between the mountain chains.

To hold the oil in a trap, the seal must be of impermeable rock. One such rock is gypsum, formed by evaporation of sea water. It is a dihydrate of calcium sulfate. As gypsum is buried deeper, it loses its water molecules and acquires the name anhydrite. Such anhydrite layers are excellent seals. The rock which holds the oil in its pores is called reservoir rock. If it is also permeable, which means that the pores are connected, oil can flow through it. The reservoir rocks in North America tend to be sandstone and those in the Middle East dolomite. Most of world's oil has been found in these anticline traps ....

Two basins that figure prominently in oil production are the Persian Gulf and the Gulf of Mexico. The newest oil provinces are the deepwater regions off the coast of Brazil and West-Africa, in the waters of Nigeria, Equatorial Guinea, and Angola. They are ancient sedimentary basins between South America and Africa ...

... the term petroleum system is used to describe the ... dynamic setting in which oil is present in some region of the world. This includes how the oil was formed, when it was formed, the structural geology of the region, and other relevant information to characterize it fully. The United States Geological Survey, USGS, ... analyzed 159 such petroleum systems in the world. This study gives petroleum geologists a comprehensive view of the world's oil and natural gas reserves, where they are, how much there are, and which parts of the world are barren of oil. It also gives them a document against which to compare their own knowledge and ... judgements on the potential for future increases in oil supply.

In summary, oil is found in select regions of the world and ... a number of conditions must be met for oil to form and be trapped. ...That these (simultaneous) conditions are often not ... met ... accounts for the large number of dry wells that oil companies drill around the globe. ... Random drilling will not find new oil. ...

Proven Reserves and Liquids

The initial estimate of oil in a field is called a scout estimate, given by a company with this specialty, or by the company holding the concession. It consists of an estimate of oil-in-place and recoverable oil. This information ... can be assigned to ...reserves. More complete seismic surveys (are then used to) modify the initial estimate and confidence in its value increases after exploratory drilling has been carried out. As a field goes into production its ultimate yield finally emerges from the rate at which its production declines. When the field is abandoned part of the oil-in-place is still left, ...

The size of the reserve is called either proven, or proven and probable, or proven, probable and possible, depending on the certainty of the estimate. Proven reserves are those which have ninety five percent chance of being produced and are therefore at the lower end of the estimated reserves. The larger estimate for proven and probable (has) ... a fifty-fifty chance of being produced. For proven, probable and possible, this drops to five percent.

The rules imposed by the Securities and Exchange Commission, SEC, in the United States requires that only proven reserves are reported. Hence, as companies develop their fields and thereby obtain a better assessment of how much of the oil-in-place can be extracted, part of what were proven and probable reserves, become proven and this part is then reported to SEC and published. This accounts for reserve growth. It also underscores how (the lack of uniform reporting) among various countries makes the task of determining what reserves to count quite difficult. ...

There is a further cause of confusion which stems from what to count as oil. Some of the hydrocarbons at high pressure in reservoirs are in gaseous state. When the hydrocarbons flow from a well, their pressure is released and some of the gases condense. This part is called lease condensate. In addition, the heavier hydrocarbons from natural gas production are in liquid form. These natural gas liquids, NGL, may also be added to the oil stream. When these and all other liquids such as refinery gains and synthetic liquids are combined total liquids production is reported. For the world oil production in 2000 was 68.2 million barrels per day and 76.7 million barrels per day for all liquids....

Estimates of World's Oil Endowment

Hubbert made two estimates of when the world oil production will peak. For the world he assumed an endowment of either 1350 Bbl or 2100 Bbl. The latter gave a peak around the year 2000. To estimate the peak production, Hubbert needed a good estimate of the ultimate oil endowment. ... in the 1950's estimates of the world endowment had increased to about 2000 Bbl. They have changed little since then. .... From this estimate (Hubbert calculated)... the maximum daily production for the world, ...(would be) ...74.0 Mb/d, which is very close to the production of 68.2 Mb/d for oil and 76.7 Mb/d for all liquids for the year 2000. This determination ... was made by Hubbert nearly fifty years ago. ...

...in a critical examination of past predictions ... those carried out in the 1970's and which were sponsored either by large oil companies, such as Esso and Shell; by governments such as U.K. Department of Energy and the U.S. Congress; or by institutions such as the World Bank, all show that world oil production would peak around year 2000. It appears that world leaders and population at large have ignored this fact ....

Country Statistics
... There are over 4000 producing oilfields in the world. .... Over 70% of world's oil comes from fields discovered before 1970. Many are in secondary recovery. .... bringing smaller fields into production will not replace the decline in these aging fields and the world is faced with a permanent scarcity. ...

Saudi Arabia has the largest reserves and production in the world. Its Ghawar field is by far the world's largest oil field with an estimated 70 GB of reserves and is believed to account for 50% of Saudi production. ... Saudi Arabia is engaged in an aggressive drilling program to keep the Ghawar field producing at its current rate. This suggests that it has entered its secondary recovery stage. ...

Iran is a very old oil province, ...Its peak production of 6 million barrels per day took place in 1974. ... Old fields, which are all in secondary recovery, account for 85% of Iran's oil production. Six of them at one time produced over a million barrels each, but now they collectively have a daily production of just over million barrels. There are new off-shore fields under development.

Venezuela is also an old oil province. Its four oldest fields produced at one time together over 2 million barrels a day, but this has dropped to 850,000 barrels. It struggles to keep its production up. ...

Iraq's oil production comes mainly from two large fields. Kirkuk, discovered in 1927 and which thirty years ago was the second largest producing field in the world. In 1990 its reserves were estimated to be 10 GB and production 1 Mb/d. Today it produces 900,000 b/d. Iraq's second largest field is Rumaila, near the Kuwait border. Discovered in 1953, in 1990 it had 11 GB left. It produces 1.2 Mb/d. These two fields account for 78% of Iraq's production. It has the most oil production potential in the world, with three 10 GB fields found since 1975, but of these only the West Qurna field is under production. Owing to the UN sanctions Iraq is likely to be the last major producer to reach its peak production. ...

United Arab Emirates has Abu Dhabi and Dubai as its two main producers, the former accounting for 85% of its production. ... Its peak production is still in the future, probably in 2008....

Kuwait's Burgan field is the second largest in the world. It was discovered in 1938 and still produces most of Kuwait's oil. Of the seven other producing fields in Kuwait, (only) two produce more than 100,000 barrels per day.

Algeria's largest field Hassi Meassaoud was discovered in 1956 and it accounts for nearly 40% of this country's production. Algeria is past its peak.

Indonesia's population of 150 million makes it one of the most populous countries in the world. It uses 70% of its oil production for domestic use. Its fields are old and...past ...peak production.

Qatar is a large producer of natural gas. Natural gas liquids and condensate add to its liquid production, as 80 % of its oil comes from a 60-year old field.

Russia's oil industry went into a downturn during the collapse of the Soviet Union. It has recovered since and now competes with Saudi Arabia for the position of the world's top producer. According to Laherrère its actual reserves are about 170 GB, hence much larger than those tabulated by Oil & Gas Journal. Still, they do not compare to Saudi reserves and the battle for oil production leadership will be short lived. Many of its fields are old and they deteriorated during the 1990's. ...

China imports more oil than it produces. China's Daqing field was discovered in 1959 and is still the fourth largest producing field of the world with over million barrels a day. Shengli field is the second largest field in China, but it is much smaller .... China's (largest fields) are now in the decline phase. ...

Mexico is one of the main exporters of oil to the United States. Its Cantarell field was discovered in 1976 and it produces over a million barrels per day. It is the third largest producing field in the world. A large gas injection to keep its production intact has been underway. ... It has unexplored territory in the Gulf of Mexico.

Laherrère, the foremost oil analyst of the world, ..(shows that) world reserves are now decreasing, ...(if we assume a) ... 2200 GB oil endowment, (it peaks in)... 2008; (if we assume an endowment of)...2100 GB, (it peaks in) 2006.

.... His analysis (also) shows that the natural gas peak will come about 20 years after the oil peak.

During the next couple of years there will be a drilling boom in the deepwater Gulf of Mexico, Gulf of Guinea, and Campos Basin off the coast of Brazil, but as production wanes from these provinces, the only oil regions left are the ultra-deep oceans, the polar regions, and the South China Sea. It is unlikely that there is much oil in Antarctica, as the crustal plates have moved the continents northward. Also the hydrocarbons in the polar regions appear to be in the form of gas. ...

There are deposits of heavy oil in the Orinoco region of Venezuela and in Alberta in Canada. The Orinoco operations are already producing more than 200,000 barrels a day, and output is expected to triple by 2006. Canada produces 400,000 barrels per day. Heavy oil production is more like a mining operation rather than traditional oil production. For this reason it is unlikely that heavy oil will be produced rapidly. Besides, in Canada the energy to produce these oils comes from stranded natural gas, stranded because there are no pipelines to bring it to markets. When the stranded gas is depleted or piped to market, energy to produce this oil must come from the final product itself, cutting the economies and certainly the energy returned for the energy invested.

Human Predicament

Although the oil industry is not yet willing to discuss frankly the imminent peak in oil production, a few industry insiders have made statements to this effect. Franco Barnabe, the CEO of ENI SpA announced in 1998 that the end of the oil era was in sight. The following year Mike Bowlin, the CEO of ARCO said that the oil industry was in its twilight. British Petroleum changed its logo during the summer of 2000 and now BP stands for Beyond Petroleum. Shell announced during the fall of 2001 that is was getting ready for the end of the oil era. ...

The issue is no longer how to solve the coming energy crisis, but how to cope with it. ..."

BP

Source: BP Statistical Review of World Energy, June 2002
Extract

Oil

There is no practical substitute for oil for some uses. It remains by far the most efficient for many others, so oil is by far the largest component of energy supply, maintaining a 40% market share, despite often being more expensive on an energy-equivalent basis than natural gas.

Crude oil is essentially a global market with modest inefficiencies caused by transport costs between the Middle East or the North Sea and the US eastern seaboard, but the cost of transporting a tanker-load of petrol or diesel fuel 200 miles means that the retail market, and hence refining, is a vast network of local markets with weak links between them.

Crude oils from different sources vary widely as to the relative amounts of the different fractions and the content of various adulterants, particularly sulfur, so the range of products produced from them by primary distillation will vary and -

> most refineries are designed to use certain varieties of crude oil and it is impossible or wastefully expensive to use certain other varieties

Although this means that in most cases one cannot directly substitute one particular cargo of crude for another, indirect substitution is usually possible as was demonstrated during the Arab oil embargo in 1973 when the seven major oil companies managed to redirect the flow of oil around the world so that no country suffered disproportionately without breaching the embargo.

Reserves

World reserves of primary energy are certainly adequate for the short-term and probably for the foreseeable future. It is difficult to be more precise since estimates are continually changing as on the one hand reserves are consumed and on the other new discoveries are made or improved technology turns into proven reserves deposits that were previously excluded because they could not be economically exploited. ...

Concerns about reserve sufficiency are more related to political worries as OPEC and FSU countries between them control over 80% of proven reserves of oil and gas. At end 2001 OPEC had 78% of oil and 45% of gas reserves and FSU had 6% of oil and 36% of gas.

Prices and Politics

OPEC countries control three-quarters of the world's proven oil reserves but account for only 40% of production.

The move by OPEC to set a higher oil price was initially totally successful. The resulting supply shortage and high prices inspired sanction for Alaskan oil production and a rush to explore for oil in politically safe areas, even if its production cost was vastly greater than Middle East oil.

Apart from the higher price and lower supply of crude caused by OPEC's intervention, retail prices include sales taxes which vary widely from state to state - UK prices for petrol and diesel are, at the time of writing, over four times as high as US prices - and distort the worldwide pattern of demand. North American consumption per head is twice that of Europe which has a comparable (albeit lower) income and GDP per head.

Energy

The oil and gas industry is fiercely competitive except in those areas where there are state-owned or state–controlled monopolies. However it is possible to make good profits upstream, since the market-clearing price of crude oil is normally significantly higher than production costs, and in those downstream markets where one supplier has a competitive advantage over its rivals.

Chemicals and lubricants between them make up less than 10% of turnover and profit so, on a simplistic view, BP's marketplace is the world market for energy. This is a long-term growth market with demand expected to rise at an average 2% (CAGR). Demand for natural gas has grown faster and it has won market share from coal.

Demand for energy tends to grow with GDP, in fact it tended to grow almost twice as fast as GDP until the early 1970s when the "Club of Rome" forecast of exhausting world energy reserves and the quintupling of the price of crude oil prompted a major drive to reduce waste and improve energy-efficiency in OECD countries.

Nevertheless world demand continued to grow roughly in line with GDP until the collapse of the former Soviet Union ('FSU') (2.3% pa compound 1975-1990). The drop in energy consumption per head in the FSU from 75% of American levels to just above European levels has offset growth elsewhere, so world consumption growth slowed in the 1990s to 1.44% pa. The underlying rate was more than 1% higher at 2.52% if you exclude the FSU and its former satellites ... .

Energy consumption in FSU reached its nadir in 1998 and has increased modestly each year since then, so projections of long term growth in world consumption at 2% pa seem reasonable. ....

The five major primary energy sources are, in order of importance:

- oil
- coal
- natural gas
- nuclear and
- hydroelectric

Fossil fuels contribute nearly 90% of the total. A number of other sources, geothermal, wind power, solar cells, are of interest but do not contribute significant amounts of energy.

Primary Energy	1991		2001	
	mtoe		mtoe	
Oil	3,138	38.5%	3,511	38.5%
Natural Gas	1,806	22.2%	2,164	23.7%
Coal	2,218	27.2%	2,255	24.7%
Nuclear	475	5.8%	601	6.6%
Hydroelectric	511	6.3%	595	6.5%
Total	8,148		9,126	

Cato Institute

August 14, 2000
Left, Right and Wrong on Energy
by Jerry Taylor, *Director of Natural Resources Studies at the Cato Institute. Used with permission.*

Rising gasoline and oil prices ... have resurrected America's perennial obsession with energy independence, the alleged remedy for OPEC production cutbacks.

Vice President Al Gore promised ... with only a few billion dollars more in subsidies, renewable-energy and energy-efficiency investments would deliver us from our oil addiction. George W. Bush and the Republican congressional leadership counter that only by increasing domestic production and reviving the nuclear option can America liberate itself from OPEC's grip. Both are peddling snake oil.

First, consider renewable energy. The federal government has spent about $ 12 billion since 1978 subsidizing wind, solar, hydro and geothermal power, yet those energy sources have managed to capture only about 2 percent of the electricity market. Efforts to reengineer the car to operate on batteries or other fuels have been a spectacular bust as well. Given that renewable energy is still several times more expensive than conventional energy, there's little reason to believe that a few billion more in government handouts will make any difference.

Nor has energy conservation turned out to be the silver bullet that its promoters advertised. Our economy is indeed more energy-efficient than it was in the mid-1970s, but oil imports continue to rise anyway. In fact, energy efficiency is a double-edged sword. As new technology makes it cheaper to use energy, we use more energy. Increasing the miles per gallon that we get from our automobiles makes it cheaper to drive and, accordingly, we drive more. Similarly, the cheaper it is to run our air conditioners, the more we run them. Nonetheless, we've spent $

30 billion of federal and state tax dollars in the past 20 years to subsidize energy conservation in an effort to repeal the law of supply and demand. ...

The plans of conservatives, however, are no better. While environmental regulations have increased the cost of domestic production and environmentalists have kept the oil industry away from some attractive oil fields, that doesn't really explain rising imports. ...

... both the left and the right have long-standing policy agendas that have failed to capture the public's imagination. You can like or dislike those agendas, but you can't make a case that either will do anything to reduce our vulnerability to OPEC production decisions. ...

The Debate Over Reserves

From: Worldenergy.org[41]

"During the 1990's, the debate over oil reserves/resources generated controversy between the "pessimists" and the "optimists".

The "pessimists" advocate the position that the world is finite and so are its recoverable oil resources. To make their argument, they rely on descriptive statistics and base their conclusions on the statistical study of past discoveries, considering all oil and gas fields to be static objects (with no evolution in the size of initially recoverable reserves). The pessimists believe that all of the oil-bearing regions worth exploring have already been explored and that the big fields have already been discovered, ergo future discoveries will be small. They claim that the official figures for proven reserves have been overestimated for some regions and that world oil production is currently at its optimum - or can be expected to reach its optimum in the medium term - and will decrease steadily thereafter.

The "optimists" hold a dynamic concept of reserves and believe that a method based solely on applying descriptive statistics to past discoveries will only yield a partial image of actual potential. The volumes of exploitable oil and gas are closely correlated to technological advances, technical costs and the price of the barrel of crude or the cubic meter of gas. For example, it is estimated that today only 35% to 40% of the oil present in discovered fields is recovered. According to an optimist, any improvement in this recovery rate - even if by only one point - allows the industry to tap substantial additional reserves. Similarly, the boundary between conventional and non-conventional hydrocarbons is not fixed, but has continued to shift regularly over time. For instance, optimists note that it is now both feasible and profitable to exploit fields at water depths exceeding 1,000 meters, which was ... thought to be impossible 15 years ago.

---

41 "Used by permission of World Energy Council, London, www.worldenergy.org"

Some of the arguments advanced by pessimists are supported by the fact that fewer "giant" fields (with ultimate recoverable reserves exceeding 500 million Bls for oil and 3 tcf for gas) are being discovered. In the 1960's, about one hundred such fields were found, but only about thirty came to light in the 1980's.

.... In 1978, the greatest production depth was 300 m. By 1998, deepwater production was under way at 1,800 m, a record set by Petrobras in the Campos Basin in Brazil. During this twenty-year period, the deep offshore sector continued to push back its technological limits.

Today, the potential represented by deep offshore resources has not yet been clearly determined. Sedimentary areas lying in over 200 m of water represent nearly 55 million km² of sedimentary basins, or four times the conventional offshore surface area. The permits that have already been delivered only cover 5% of this area.

For oil companies, the next target depth is 3 000 m. Meeting this objective constitutes a major industry challenge for the next 5 to 10 years."

ExxonMobile

(Author's note: There are some very optimistic thinkers that believe we humans can find an oil bonanza by drilling wells into sediments that lie below the ocean at depths of more than 500 meters. Here is what we will encounter.)

How deep is deep?

Since drillers first moved offshore more than 50 years ago, the meaning of deep water has changed. To many in the business, it means water too deep for conventional freestanding steel platforms. Today that depth is roughly 400 meters (1,300 feet), greater than the height of the Empire State Building in New York City. The industry's definition of deep water will change again as ExxonMobil and others begin exploring in water depths of 3,000 meters (10,000 feet) and more.

As a leading explorer of deepwater basins, ExxonMobil holds interests in more than 135 million gross acres in waters deeper than 400 meters (1,300 feet). We participated in over 30 major discoveries and are the largest multinational holder of deepwater acreage in three of the world's most active regions: West Africa, Brazil and the Gulf of Mexico.

Exploration

Three-dimensional imaging, a powerful tool for finding oil and gas, uses seismic data and the world's fastest computers to create images of deepwater basins. ExxonMobil pioneered many of the techniques for gathering seismic data and wrote the industry's most sophisticated software to process and interpret it. Proprietary 3-D imaging is one reason that in recent years, more than half our exploration wells find oil.

Drilling

With great precision, drillers now aim their bits at targets that are miles from the platform. Known as extended reach drilling, this technology helps recover oil and gas from a wide area using a minimum of offshore platforms. Related technology allowed ExxonMobil to drill horizontal wells at record water depths in the Hoover-Diana field.

Development

Deepwater wells are significant capital investments, so the number of wells drilled and how widely they are spaced often means the difference between a field's economic success and failure. Much of ExxonMobil's research targets new ways to make wells produce more oil and gas at lower costs. ...

What changes with water depth?

Like spacecraft, the equipment we use offshore must operate flawlessly for decades in hostile environments where much of the work is done by remote control. Our success in exploring and developing deepwater oil and gas prospects comes from our ability to deal with the sea's many challenges, including:

- Pressure. A mile deep, water squeezes everything at more than one ton per square inch. Pressure is a major factor in our designs for pipelines and subsea equipment.
- Temperature. Below 600 meters — anywhere in the world — the ocean is a steady 4 degrees Celsius (39 degrees Fahrenheit). On some deepwater gas wells, these low temperatures can cause water vapor and natural gas to form ice-like crystals, blocking the flow through pipelines.
- Marine life. Colonies of worms and mussels often thrive around naturally occurring oil and gas seeps. Over thousands of years, their remains have formed rock-hard deposits that must be avoided to prevent damage to equipment on the ocean floor.
- The seabed. In some places, the ocean floor is very soft, and any unsupported equipment will sink out of reach. Elsewhere, underwater hills and valleys pose the threat of sediment and rock slides that can damage subsea wells.
- Currents and waves. Currents can complicate the installation and operation of offshore equipment. Storms can generate waves taller than a seven-story building and wave crests moving at 20 knots.

Matt Simmons

Mike Ruppert of From the Wilderness interviews energy investment banker Matt Simmons on the future of energy in America. Thank you Mike for allowing me to use this excellent interview.

Interview abstract. September 20, 2003

FTW: What's the most important thing you want the American people to know about Black Thursday (the day of the Northeast and Midwest blackout) ?

SIMMONS: This blackout ought to be ... telling us about a host of energy problems that are ultimately going to prevent any future economic growth. ... The event itself was astonishing. Senior people like ... the head of NERC [North American Electric Reliability Council] were asking how this could happen. But the problem was inevitable. The only thing we didn't know was when it would happen.

FTW: What did happen?

SIMMONS: ... what happened was deregulation.

Deregulation destroyed excess capacity.

Under deregulation, excess capacity was labeled as "massive glut" and removed from the system to cut costs and increase profits. Experience has taught us that weather is the chief culprit in events like this. The system needs to be designed for a 100-year cyclical event of peak demand. If you don't prepare for this, you are asking for a massive blackout. New plants generally aren't built unless they are mandated (and supported by regulation), and free markets don't make investments (in new plants if they) give one percent returns. There was also no investment in new transmission lines.

... every aspect of carrying capacity, from generators, to transmission lines, to the lines to and inside your house, has a rated capacity of x. When you exceed x, the lines melt. That's why we have fuse boxes and why power grids shut down. ... Another problem was that with deregulation, people thought that they could borrow from their neighbor. New York thought it could borrow from Vermont. Ohio thought that it could borrow from Michigan, etc. That works, but only up to the point where everyone needs to borrow at once and there's no place to go. ...

FTW: And natural gas too?

Oil, Jihad and Destiny

SIMMONS: ... people need to understand the concept of peaking and irreversible decline. It's a sharper issue with gas, which doesn't follow a bell curve but tends to fall off a cliff. There will always be oil and gas in the ground, even a million years from now. The question is, will you ... (be willing to) spend hundreds of thousands to drill a gas well that will run dry in a few months? ... There aren't going to be any dramatic new discoveries and the discovery trends have made this abundantly clear.

We are now in a box we should never have gotten into and it has very serious implications. We also see the inevitable issues that follow a major blackout: no water, no sewage, no gasoline. The gasoline issue is very important. Our gasoline stocks are at near all time lows. With the blackout, more than seven hundred thousand barrels per day of refinery capacity were shut down. People were told to boil their water. (But) ... they go to their electric stove which isn't working. What then? ...

FTW: So how big a factor was the weather?

SIMMONS: It was THE factor in my opinion. To show much weather determines power use, in the week of August 3rd, the US set an all-time national record for electricity use of 90,000 Gigawatts. The Mid-Atlantic States' use of power had jumped 29.5% over last year and 20% over just the previous four weeks. Why? The temperature had been as hot as we experienced on Black Thursday. ... Everything that happened on August 14 started in the 17th hour. (5 PM at various local times). That's when everything is running at once: industrial, residential, and commercial. This is when demand peaks regardless of the weather. And we know that in hour 17 on that day the US experienced all-time peak energy use. That's when the system tripped out.

FTW: So we have two basic camps saying that the problems are generating capacity and transmission lines... What about the advocates for deregulation who argued that there would be more generating capacity as a result?

SIMMONS: History answers that one. Following the 1965 blackout when NERC was created there was a mandate (by regulatory agencies) that publicly owned and regulated power providers had to build new plants. Every five years, ten per cent was added to the generating base. As deregulation was implemented in the 1990s, it was argued that it would open up vast quantities of energy in neighboring states. In the first five years of the decade, only four per cent capacity was added (by a deregulated industry) over the entire period. In the second five years, only two per cent was added.

(Then) in the summer of 1999, we had thirty consecutive power events which unleashed the single biggest construction boom in history which built 220 thousand megawatts of new plants at a capitalization cost of six to seven hundred thousand dollars per megawatt. Ninety-eight per cent of those plants were gas fired.

It was decided to use solely natural gas plants for several reasons. Coal fired plants took five to seven years to build. They are very dirty environmentally and the permit process is difficult. We have built on all the available hydroelectric sites we can build on. Nuclear is unpopular and expensive. Oil fired plants are remnants of the days when oil was cheap. Those days are not coming back because Peak Oil is with us now. Besides that, oil fired power plants are about the least efficient use of a barrel of oil that I can imagine. That left natural gas and the economists mistakenly presumed there would be large supplies.

> But natural gas (electricity generation) plants were built with no supplies. Synthetic contracts were used, Enron-style, to sell gas futures when the gas didn't necessarily exist.

FTW: Assuming that there was enough feed stock to run the new plants how much building are we talking about?

SIMMONS: Each state would need to build forty to fifty per cent excess capacity. A forty per cent cushion merely provides the chance to withstand a day of high summer heat and the chance to grow by about 3% per year for three years.

FTW: Yet even if we re-regulate there are still going to be problems with feed stock to power the plants. How serious is that?

SIMMONS: Someone's going to be left holding the bag big time. If natural gas consumption surges in ten days of excessive heat then it would require almost a complete shutdown of industrial consumption to ... protect the grid. ...

> there isn't going to be enough (natural) gas to run those (electric) plants,
> ....

FTW: You mean shut down the economy for ten days to keep people from cooking?

SIMMONS: Yes. ...

FTW: What about imports of natural gas from overseas? Russia and Indonesia have huge reserves; Canada, as the Canadians are painfully aware, is almost depleted when it comes to natural gas.

SIMMONS: Indonesia's gas fields are very old. Its Natuna gas fields, a source of stranded gas that gets discussed all the time has 95% CO_2 and apparently costs about $40 billion to develop a mere 1 bcf/day of dry gas. Russia has four old fields that make up over 80% of their gas supply and they all are in decline. Canada's decline problems are as serious as the US.

FTW: Windmills? Solar?

SIMMONS: There's no way they can replace even a portion of hydrocarbon energy.

FTW: Reducing consumption?

SIMMONS: Reducing consumption has to happen, but many of the favorite conservation concepts make little overall difference. The big conservation changes end up being steps, like a ban on using electricity to either heat water or melt metals and instead, always using (natural gas for heating and cooking in the consumer's home). The latter is vastly more efficient. The energy savings are enormous. We (also) need lower ceilings and smaller rooms. We need mass transit, and to eliminate traffic congestion. Finally, we need a way to keep people from using air-conditioning when the weather gets really muggy and hot at same time. The strain this puts on our grid is too overwhelming. We also must begin to use our current discretionary power during the nighttime. All of theses steps are hard to implement but they make a difference.

FTW: What is the solution?

SIMMONS: ... The solution is to pray. Pray for mild weather .... Pray for no hurricanes and (no reduction of) natural gas supplies. Under the best of circumstances, if all prayers are answered there will be no crisis for maybe two years. After that it's a certainty.

FTW: On that cheery note let's take a look at oil supplies.

SIMMONS: ... Last month the IEA (International Energy Agency) updated their database. They had for years been talking about a coming huge surge in non-OPEC supply, ... It hasn't happened.

We have the highest oil prices in 20 years
and ... great technological advances (but these) have not
had a measurable impact on discovery or production.

FTW: I have recently noted the speed with which the Chad-Cameroon pipeline was built and switched on. Chad only has estimated reserves of around 900 million barrels (World consumption is I billion barrels every 12 days). I see a sense of urgency there.

SIMMONS: It's amazing. What's that pipeline going to pump, fifty thousand barrels per day? That figure may go up, but it's inconsequential in the long run. It's a sign of how strapped world supplies really are and that we may be finding out that we are already over the peak.

176

FTW: What about Iraq and Saudi Arabia? We have been following Iraq closely and all the sabotage, infrastructure damage and the pipeline bombings are actually reducing Iraqi capacity. That leaves Saudi Arabia with 25% of known reserves.

SIMMONS: I have for years described two camps: the economists who told us that technology would always produce new supply and the pessimists ... who told us that peak was coming in maybe fifteen or twenty years. We may be finding out that we went over the peak in 2000. That makes both camps wrong.

Over the last year. I have obtained and closely examined more than 100 very technical production reports from Saudi Arabia. What I glean from examining the data is that *it is very likely that Saudi Arabia, ... has very likely gone over its Peak. If that is true, then it is a certainty that planet earth has passed its peak of production.* What that means, in the starkest possible terms, is that we are no longer going to be able to (increase our oil production) ...

FTW: What about people like Alan Greenspan and popular writers who tell us that there is no basic problem with energy supplies? Others offer us hydrogen, which is laughed out of hand by people who have looked at its feasibility and efficiency.

SIMMONS: Basically they just don't get it. Some of them have gotten lazy. They were so carried away by the arguments of the economists that they stopped doing their homework. Month by month, and year by year, events are proving them systematically and thoroughly incorrect. They just don't get it. Right now, there is a deluge of stories on the wonders of hydrogen. This is another area of great confusion.

Hydrogen is not a primary source of energy.
For a Hydrogen Era to occur you need
an abundance of natural gas, or you need to create (hundreds) ...
of new power plants using coal and nuclear power. ...

FTW: But peak oil is peak oil, is it not? Aren't we just talking about something that would have delayed the inevitable for a few years? ...

SIMMONS: Peaking of oil and gas will occur, if it has not already happened, and we will never know when the event has happened until we see it "in our rear view mirrors."

Mike Ruppert is a former Los Angeles police officer and investigative journalist. He publishes a monthly newsletter and sponsors a web site at. www.fromthewilderness.com, and www.copvcia.com

Matthew Simmons is the CEO of the world's largest Energy Investment Bank, Simmons & Company International. It has a web site located at (http://www.simmonsco-intl.com/). Its clients include Halliburton; Baker, Botts, LLP; Dynegy; Kerr-McGee; and the World Bank.

"Reprinted with permission, Michael C. Ruppert and From The Wilderness Publications, www.copvcia.com, P.O. Box 6061-350, Sherman Oaks, CA, 91413. 818-788-8791. *FTW* is published monthly, annual subscriptions are $50 per year."

Myths and Realities of Mineral Resources

Extract from Chapter 27 of GeoDestinies: The inevitable control of Earth Resources over nations and individuals, Walter Youngquist Ph.D., Professor of Geology, University of Oregon; National Book Company (1997). Extract used with permission.

Although minerals and energy minerals are fundamental to our existence, the facts of these resources and of industries which produce these materials are subject to many myths and much misinformation. This is unfortunate for it clouds the ability of individuals in a democracy to make intelligent choices. Some of the distortions are deliberately made by political interests who play upon the fears and hopes of the electorate, and then in the role of the defender of the public interest against the oil or mining companies seek to obtain votes by this device. Some statements are made from ignorance, and some are made by people who have their own political and social agendas which they wish to perpetrate upon the public. Some are made by people who are a bit over enthusiastic about a particular resource and do not carefully examine the hard facts, or may not be aware of them. Some statements are made by promoters wanting to raise money for a particular mineral development, whether that development has a sound basis or not.

It is important that facts be sorted out from fiction ....

Myth: There is no oil supply problem in the United States

During the two oil supply crises of 1973 and 1979, in the U.S. the average citizen frequently stated the belief that no "real" oil shortage existed, and that the shortages were caused by the oil companies withholding oil from the market...

Reality: The United States passed the point of oil self-sufficiency in 1970, and has been an importer of oil ever since then. ....The United States is the most thoroughly drilled area in the world and there is no possibility that this nation will ever again be self-sufficient in oil ....

Myth: Oil companies have capped producing wells to keep up the price of oil

This is one of the oldest and most persistent myths about the oil industry. The idea is that oil companies will drill wells and then cap them, thus withholding production from the market until the price of oil goes up.

Reality: It is true that many wells are drilled and then capped. Almost all of them are capped because they are dry holes—that is, they are failures. ....There are some wells which could produce oil which are temporarily capped. There are two common reasons for this. One is that there is no facility for transporting the oil from the well at the moment. Either a pipeline does not exist or it is too expensive to truck it out. Generally, if the well is a producer, other wells will be drilled in the area to establish the presence of enough recoverable oil to justify developing a transport system by which the oil can be brought out economically.

A second reason may be that occasionally it is true a well may be drilled, completed, and capped when the current price of oil is not high enough to pay for the expenses of producing the oil—the pumping costs and perhaps the problem of the disposal of the salt water which may be produced with the oil. However, capping a well and leaving it for a time is risky because sometimes the well cannot be restored to production.

Drilling a well is so costly that if the well is productive and capable of bringing a return on investment, the well will be produced.. ....

Myth: Don't drill this prospective field. Only 90 days of U.S. oil supply there

One of the most misleading arguments used against drilling a particular area is the statement that it would only supply X number of days or months of U.S. oil demand. .... It is one of the most widely and most effectively used arguments against oil drilling. ....

Reality: At the present time the U.S. uses about 18 million barrels of oil a day (1996). A 100 million barrel oil field is regarded in the petroleum industry as a "giant." They have been discovered only infrequently. Yet if one of these giant oil fields was used to supply U.S. oil demand, it would last less than six days!

To put this in further perspective, ... These days, a ten million barrel oil field discovery is an important event in U.S. oil exploration. But that amount would last the U.S. less than 14 hours! The fact is, we are not discovering ten million barrel oil fields every 14 hours in the U.S. That is why our oil reserves are in decline. Prudhoe Bay, the largest oil field ever discovered in North America, would have lasted the U.S. less than two years if it alone had been used.

But it is not possible to produce all the oil out of Prudhoe Bay or in any other field in 90 days, or six months or two years. If one divides the number of producing oil wells in the U.S. into the total proven U.S. reserves, each well has a reserve of about 38,500 barrels. These 38,500 barrels of oil, if they could be immediately produced, would supply U.S. oil demand for about three minutes. On this basis, it might be argued that none of these wells should have been drilled, in which case the U.S. would have no oil production. But oil supplies are produced over many years from many wells which make up the total U.S. production.

Each well makes a contribution, ... Those who would curtail exploration first need to reflect on what is causing the huge and increasing demand on mineral and energy resources, and address that cause and not the symptoms of the problem. ...

Myth: oil companies own oil

Reality: In a number of countries, including Saudi Arabia, Venezuela, Kuwait, Iran, Iraq, Peru, and Mexico, oil was originally discovered and developed by foreign companies with the expertise which the country itself did not have. Subsequently, with the rising tide of nationalism following the colonial period, oil company properties—oil fields, pipelines, shipping facilities—were taken over by the respective governments, at times with little or no compensation.

Most of the oil in foreign countries is owned by the governments, not the oil companies. Oil companies simply hold leases (abroad commonly called concessions) to develop the oil deposits. .... In a sense they own the oil they produce, but they never really own the oil in the ground. They only lease the right to produce it. This is an important point, because it means that U.S. companies or any other companies operating in a foreign country do not own an assured safe resource base.

In the United States, the mineral rights which include oil and gas usually belong to the owner of the land. ... Offshore oil belongs either to the adjacent state, or beyond the state limits, to the federal government. Oil companies, ... pay a royalty to the private owner, or royalties and taxes to the government. These costs range from 12.5 percent to as much as 90 percent of the value of the oil....

Myth: Oil companies make big profits compared with other enterprises

The profits of oil companies are frequent targets of criticism by both the politicians and the media. ....

Reality: ... the amount of capital which has to be invested in the production of oil is very large and it takes a long time, in some cases, many years, before any return can be realized on the investment, if indeed there is a return at all. ....

Oil exploration and production is a high risk venture. Companies that do survive, earn a relatively modest return on investment. On records kept since 1968, the average return on stockholder investment in 30 representative U.S. oil companies has been 12.5 percent. In 1994, it was only 9.2 percent. For 30 representative manufacturing companies, the return has been 13.1 percent. The average return for oil companies is less than the average return for manufacturing industry in general....

Myth: Alternative energy sources can readily replace oil

This is the assumption made by many people who advocate alternative energy sources as an early easy solution to our dependence on imported oil, and the perceived negative environmental effects of burning oil.

Reality: The facts relative to this myth are mixed. Alternative energy sources can replace oil in its energy uses, but in some uses much less conveniently than in others. Fuel oil used under steam boilers can be replaced by nuclear fuel, or coal. But replacing gasoline, kerosene, and diesel fuel for use in vehicles, airplanes in particular, by an alternative energy source will be much more difficult.. At the present time, 97 percent of the world's approximately 600 million vehicles are powered by some form of oil. Going to another fuel source to meet this huge energy demand now met by the convenient, easily transported, very high grade energy source which is oil will not be easy....

Myth: Alternative energy sources can simply be plugged into our present economic system and lifestyle, and things will go on as usual....

People do not appreciate the close relationship between the current energy sources, principally oil, and the control which energy forms have over the activities of their daily lives, and where and in what sorts of structures they live and work, and use for transportation.

Reality: ... Other energy sources, beyond oil, ... would involve a restructuring of daily routines. Our activities are very much controlled by the energy forms which we use. Our standard of living is largely a function of how much and in what form we can command energy supplies. Changing from the energy form which is oil to other energy sources can and will have to be done, but lifestyles will be altered, as may also be the standard of living.

Myth: Biomass—plants—can be a major source of liquid fuels...

Reality: ... There are several reasons why converting growing plants to oil will not be a significant substitute for oil obtained from wells. ...The energy conversion efficiencies are low, in some cases as with ethanol from corn, it is negative. The energy cost of harvesting and transporting the materials is high relative to the energy produced. ...The volumes of plant material available are not sufficient to yield large amounts of oil, given the low energy conversion efficiencies. The degradation of the land growing these materials by continuing harvesting without returning the fiber to the land is severe. ... the Energy Research Advisory Board of the U.S. Department of Energy stated in 1981 ... that 258 million Americans used 40 percent more fossil energy than the total amount of solar energy captured each year by all U.S. plant mass. .... Biomass is not a potential source of significant quantities of liquid fuel.

Myth: There are billions of barrels of oil which can be readily recovered from oil shale in the U.S.

As the United States has the world's largest and richest deposits of oil shale, the optimistic statements which sometimes arise from that fact are among the more commonly heard ...

Reality: The supposedly great prospects for the production of oil from oil shale in the United States has been one of the most widely promoted ... energy myths . .... there (has) ...been no demonstrated methods of oil recovery at costs competitive with oil of comparable quality, ... A variety of processes have been tried. All have failed. Unocal, Exxon, Occidental Petroleum, and other companies and the U.S. Bureau of Mines have made substantial efforts but with no commercial results....

Myth: Canada's oilsands with 1.7 trillion barrels of oil will be a major world oil supply. It appears to be true that in the Athabasca oilsands and nearby related heavy oil and bitumen deposits of northern Alberta there is more oil than in all of the Persian Gulf deposits put together.

Reality: The impressive figure of 1.7 trillion barrels of oil is deceiving. It is likely that only a relatively small amount of that total can be economically recovered. The oil is true crude oil but it cannot be recovered by conventional well drilling. Almost all of it is now recovered by strip mining. The overburden is removed and the oilsand is dug up and hauled to a processing plant. There the oil is removed by a water floatation process. The waste sand has to be disposed of.

Much of the oilsand is too deep to be reached by strip mining. Other methods are being tried to recover this deeper oil, but the economics are marginal. .... Canada will probably gradually increase the oil production from these deposits, but until the conventional oil of the world is largely depleted these Canadian deposits are likely to represent only a very small fraction of world production. The production will always be insignificant relative to potential demand. Oilsands are now and will be ...

Myth: ... the widespread concept remains that conservation can solve the energy problem.

Reality: Energy and mineral conservation and recycling are useful goals, but conservation is only a temporary solution to the overall problem of continued growth of energy demand from an ever-increasing population. To accommodate more and more people, each person might use less and less resources, but at some point there is a minimum amount of the resource which has to be used. .... There is no way to ultimately conserve out of the energy supply problem against an ever-increasing population. Demand can be reduced but if at the same time, an increase in population absorbs those savings, there is no gain. ....

Myth: ...—"we will achieve energy independence"... (is) a popular political campaign promise ... Citizens look for cures to their problems, and the candidate who can most convincingly promise them may be the winner.

... it may be noted that, win or lose, soon after the campaigns have been over, the goal of energy independence seems to have been lost in the shuffle of everyday politics as usual.

Reality: It may be hoped that U.S. energy independence can eventually be achieved,.... Political posturing and optimism will not solve the energy supply problem. However, political decisions can encourage (the) development of alternative energy supplies, ...

Energy independence for the United States is at present becoming less and less a near term possibility. The economy continues to be based very largely on petroleum, and oil imports continue to increase each year. Any political candidate who states that energy independence can be achieved for the United States in any presidential term of office (or even in two or three decades) is simply either not being honest or is totally ignorant of energy supply, and the prospects for viable alternatives.

A national move toward energy independence, which has to be expressed by the citizens through their elected representatives in the Congress, has not materialized. Energy independence for the U.S. will remain a myth if the present energy course is continued. ....

Myth: "At current rate of consumption ..."

This is commonly used as a comforting statement to assure the public that there is no looming shortage of a given resource. "At the current rate of consumption" a given resource will last for at least X number of years. Usually, this is quite a long time. There is no problem.

Reality: This very misleading myth is that the "current rate of consumption" does not represent the future. The rate of consumption of almost all resources, particularly energy, is increasing every year. The increase in resource consumption is caused by three factors: population growth, a demand for an increase in per capita consumption of a resource to increase living standards, and a larger number of uses found for a given resource. Oil is the classic example which illustrates increased demand from all three causes. Present demand for oil is increasing at the rate of about two percent annually, which means demand will double in 35 years. ....

Demand does not grow arithmetically, but increases exponentially. That is, it goes up as a percentage each year over the previous year. Therefore, the statement that a depletable resource will last for X number of years "at current rate of consumption" has little relation to the reality of the actual life of the resource. .... There is little, if any, possibility that the amount of oil available worldwide 40 years hence will be the same as today. It will be less, and the critical point is *when world oil production begins to decline,* not when the last drop of oil is ever pumped from the ground....

Myth: The omnipotence of science and technology—it can do anything

There continues to be a belief in some circles that technology and science can indeed solve all problems of human material existence indefinitely, ...

Reality: ....Faith that science and technology can solve all resource supply problems is evidenced by the widely expressed public view that "you scientists will think of something." It ignores the fact that something cannot be made from nothing, and in order to have a resource one must have some material thing with which to work. This fact, however, is met with the thought that substitutions can be made. This is true, within the reality that eventually substitutions also become exhausted. Also, there are definite limits as to what substitutions can be made. There is, for example, no substitute for water. ....

Science and technology do have limits imposed by the immutable laws of physics, chemistry, and mathematics. At the present time it seems clear that if current trends continue in growth of population, the demands of the human race will soon overwhelm the ability of science and technology to solve the problems of availability of resources, which are the basis for human existence....

Myth: Because past predictions of resource and population problems have proved incorrect, all future such predictions will not come true, therefore there is no need to be concerned....

Reality: Malthus—then and now Malthus' predictions were wrong because he did not foresee the coming industrial and scientific revolution. ... The problem is that science and technology will not be able to continue to discover and develop the amount of new resources

184

necessary to support a population growing at an exponential rate. ... Advanced exploration and production technologies have allowed geologists and engineers in a less than two hundred years to discover and develop the huge store of mineral and energy resources which accumulated slowly over billions of years. In a fraction of a second in terms of the length of human existence, Earth resources basic to civilization have been brought into production in volumes never before seen.

Soils, oil, high grade metal and coal deposits and now those of lower grade, groundwater, and other resources including dam sites, are being used up at an unparalleled rate. Since 1900, world population has increased nearly four times, but the world economy has expanded more than 20 times. Fossil fuel use has increased by a factor of 30 and industrial production has grown by a factor of 50, and four-fifths of these increases have occurred since 1950. Civilization exists now in a new reality which is far different from that of Malthus's time. Population grows but mineral and energy resources do not increase. By discovery and advanced recovery technology, the immediate supply can be made to increase, but in total, minerals and energy sources with the exception of sunlight, are depletable.

Since the beginning of the Industrial Revolution, the speed of human assault upon Earth's resources has greatly increased. More petroleum, coal, and metals have been used since 1950 than in all previous human history. In the United States the high grade, easily won, low cost deposits of iron ore (hematite), copper, and petroleum have been depleted. In some other regions of the world, high grade deposits still exist but are rapidly being developed and used. There are few major dam sites in the United States on which to build large reservoirs for additional hydroelectric power, and irrigation projects.

...*Promotion of Myths*

The media—newspapers, magazines, television, radio—report the news. But in the competitive haste to do so, sometimes they become accessory to spreading misinformation. The statements by uninformed people, politicians pursuing votes, unscrupulous promoters, or citizen groups trying to further a particular point of view may ignore realities. Too often these statements are picked up by the media and reported as fact....

It is perhaps too much to ask the media to thoroughly examine facts behind such statements. But there should be at least some minimal effort to do so because there is an unfortunate tendency for people not to critically read what is in the papers, or thoughtfully examine what television and radio brings them. Most do not have the background to make critical examinations. In the case of broad sweeping statements on things so vital as energy supplies, the media could at least quite quickly get a second opinion and present that also, which would give a useful balance to the reporting....

It has been said that "optimists have more fun in life, but pessimists may be right." ...Regardless of the popularity of optimism over realism, the wisest route for humanity would be that plans and decisions be based on today's scientific and technological realities and reasonably visible resources, rather than on hopes for things which may never arrive. Optimism is vital in looking toward the future. One must be optimistic as a basis for making an effort. But optimism should be tempered with facts. The media and government leaders should try to learn the facts, and then have the courage to state them. Campaigns for public of office should not lead the citizenry into false hopes. As civilization proceeds, it will be much more convenient and less disruptive to be pleasantly surprised along the way than unpleasantly surprised. Myths must be replaced by reality on which intelligent decisions are made.

"Facts do not cease to exist because they are ignored."
-- Aldous Huxley

Dr. Walter Youngquist has worked as a geologist in 70 countries. He has spent a lifetime studying the vital relationship of Earth resources to nations and individuals. His professional affiliations include membership in the American Association of Petroleum Geologists, the Geothermal Resources Council, and the New York Academy of Sciences. He is a Fellow of the Geological Society of America and the American Association for the Advancement of Science.

Appendix 2: Alternative Energy

Fuel Cells

Technology

Fuel Cells are often hailed as the future of advanced energy technology. Fuel cells, the theory goes, eliminate the pollution problems of oil combustion. Oxygen from the air is combined with stored hydrogen to produce electricity. The only by-product is water. The theory also contends that since there is an inexhaustible supply of hydrogen and oxygen, fuel cell technology will protect us from the problems of oil depletion.

Like batteries, fuel cells use electrodes (solid electrical conductors) in an electrolyte (an electrically conductive medium). Hydrogen fuel is fed into the "anode" of the fuel cell. Oxygen (or air) enters the fuel cell through the cathode. Encouraged by a catalyst, the hydrogen atom splits into a proton and an electron, which take different paths to the cathode. The proton passes through the electrolyte. The electrons create a separate current that can be utilized before they return to the cathode, where they are combined with the oxygen to form a molecule of water.

> Unfortunately, the existing storage and handling characteristics of hydrogen
> make it impractical as a replacement for oil products
> in most combustion applications.

The industrialized nations have been increasing their hydrogen power research and development efforts. There are, unfortunately, many challenges that must be overcome before hydrogen powered fuel cells are a practical reality for all combustion applications. Although there are many interesting laboratory experiments and demonstration programs, it would be a big mistake to automatically assume these projects will yield a workable fuel system.

Fuel cell power is expensive. Most of the current designs require costly precious-metal catalysts or materials that are resistant to extremely high temperatures. Fuel cell durability and reliability problems add to the cost of operation. The manufacture of hydrogen is energy intensive. Hydrogen is currently more expensive to produce than conventional fuels, such as gasoline, and many of the more cost-effective production methods generate greenhouse gases. Hydrogen transportation, distribution, storage and consumption will require the development and deployment of a costly new fuel system infrastructure.

Reliability and durability must be improved. High-temperature fuel cells are prone to material failure. Corrosive materials, such as the potassium hydroxide or phosphoric acid used in some designs, shorten fuel cell life. PEM fuel cells must have effective water management systems to operate dependably and efficiently[42]. Finally, all fuel cells are prone, in varying degrees, to catalyst poisoning, which decreases fuel cell performance and longevity.

Hydrogen has a low energy density (0.0007 pounds per gallon) versus gasoline (6.0 pounds per gallon). Hydrogen's low energy density makes it impractical for many applications because there is no practical way to store enough hydrogen in a compact space for later consumption. This is a major challenge, for example, in the use of hydrogen as a vehicle fuel. In order to fix this problem, high-pressure storage tanks are currently being developed, and research is being conducted into the use of other storage technologies such as metal hydrides and carbon nanostructures (both are materials that can absorb and retain high concentrations of hydrogen).

So we must either combine hydrogen for storage, transportation and consumption with another hydrocarbon (so it can be compressed and stabilized) or absorb the hydrogen atoms into another structure (so it will not blow up) until it is released for consumption. Reformers, which are devices that turn a hydrocarbon or alcohol fuel into hydrogen and carbon atoms, can be used for both mobile and fixed site applications. Methanol, natural gas and propane are typical reformer fuels. Unfortunately, the hydrogen that comes out of a reformer may be contaminated with the atoms of other elements, reducing the efficiency of the fuel cell. Reformers also produce other gases (such as carbon dioxide) and can generate excessive heat. Hydrogen can be derived from many hydro-carbon fuels by reforming, but this reaction imposes a reduction in practical efficiency because energy is lost in the reforming process

If we could run our fuel cells on pure hydrogen, and we did not need to manufacture the hydrogen, then fuel cells - as an energy system - would have an efficiency of roughly 80 percent. That is, 80 percent of the energy content would be converted into electricity. If we must store hydrogen in a hydrocarbon or alcohol chain, and use a reformer to liberate the hydrogen, then the efficiency of our conversion of energy to electricity drops to less than 40 percent. The mechanical operation of an electric motor, which also absorbs energy, will drop the efficiency of our fuel cell further - perhaps to less than 32 percent.

---

42 Although the current darling technology of enthusiasts, the membrane of Proton Exchange Membrane fuel cells may dry out, get too hot, become water logged, or become contaminated with trace amounts of carbon.

In automotive applications, the efficiency of a fuel cell is competitive with conventional gasoline powered engines. Because of all the pumps, fans and generators that go with a conventional internal combustion engine, and because much of the fuel energy is lost as heat, our automobiles have an efficiency of less than 25 percent. The fuel cell is also competitive with the utility power grid for the delivery of electricity. Phosphoric acid fuels cells - already available on a commercial basis - generate electricity at more than 40% efficiency and nearly 85% of the steam this fuel cell produces can be used for co-generation. This potential efficiency compares to a 35% efficiency for the utility power grid in the United States.

Unfortunately, our efficiency comparison between fuel cells and gasoline or diesel engines does not begin and end with the storage, transportation and combustion processes each one uses to covert thermal energy into mechanical energy. Hydrogen - irrespective of whether it is derived from an alcohol, a hydrocarbon, or water - must be manufactured. It is not a source of energy. Hydrogen is a carrier of energy. Thus when we calculate the efficiency of a fuel cell, we must also deduct the amount of energy it took to make the alcohol, hydrocarbon or pure gas. When we add these costs back into our efficiency equation, the hydrogen fuel cell is not economically attractive. In addition, using an oil based product for the hydrocarbon resource does nothing to reduce the impact of an oil depletion crisis.

> What we really need is a way to convert NON- PETROLEUM raw materials into a hydrocarbon chain that has masses of hydrogen atoms, can be compressed into a liquid, will not blow up during transportation and handling, does not require the excessive use of toxic materials in order to generate electricity, has a high conversion efficiency and is competitive in cost with oil at $40 - $50 per barrel.

Applications

Fuel cell electric power for portable and fixed site applications? We have a long way to go in the development of consumer and industrial products, and we need to solve the practical problems of manufacturing, storing, handling, distributing, and consuming NON-PETROLEUM hydrogen fuels before this technology can fulfill its promise. Fuel cells developed for the space program in the 1960s and 1970s were extremely expensive and impractical for terrestrial power applications. During the past three decades, significant efforts have been made to develop more practical and affordable designs for stationary power applications. But progress has been slow. Today, the most widely marketed fuel cells cost about $4,500 per kilowatt; by contrast, a diesel generator costs $800 to $1,500 per kilowatt, and a natural gas turbine can be even less.

But the commercialization of practical fuel cell technology is in progress.

For portable power applications, such as cell phones and laptop computers, Direct Methanol Fuel Cells are about to make their appearance in practical consumer products. Alternative designs are in the works for military and mobile service applications. In portable power applications, fuel cells promise to be a superior replacement for conventional batteries. Fuel cells will power these devices for a longer period of time than existing battery technology and full electrical power can be restored by simply filling the (very small) fuel cell tank with an appropriate hydrogen compound.

Electric power generation also shows promise as a fuel cell application. We may find that it is more efficient to provide electric power to rural homes and farms using locally installed fuel cells, than by generating the electricity at a huge power plant and sending it to a remote consumer through a conventional electric power grid. Portable fuel-cell systems are already available for providing backup power to hospitals and factories. The fuel-cell technologies being developed for these power plants will generate electricity directly from hydrogen and will use the heat and water produced in the cell to power steam turbines and hence generate even more electricity. Fuel cells are competitive in cost for a limited number of stationary power generation applications if one factors in all of the associated infrastructure costs that go with power generation. On the other hand, if the DOE hits its target of $400 per kWh and the price of oil increases as projected in this report, then fuel cells could become the power generation technology of choice for building, campus and small community applications by 2010 - 2015.

Press Release
World's largest clean coal fuel cell almost ready for initial tests.

FuelCell Energy, Inc., Danbury, Conn., will test a two-megawatt fuel cell system at the Wabash River Energy, Ltd., coal gasification-combined cycle power plant. Developed under the Department of Energy's Fossil Energy program, the molten carbonate fuel cell system will demonstrate a highly efficient, pollution-free electricity production system.

The objective will be to generate electricity from coal without combustion in the world's first coal-fuel cell demonstration power plant. It uses an electrochemical reaction between fuel and oxygen from the air to produce electric power. The project will use technology developed by FuelCell Energy through a research partnership that began more than 25 years ago with the Department's National Energy Technology Laboratory.

The 260-megawatt Wabash River plant has been operating since November 1995 and is currently one of only two commercial-scale coal gasification power plants running in the United States. PSI Energy's Wabash River integrated gasification combined-cycle plant converts coal into a synthetic gas as a feed stock for the fuel cell. Because the fuel cell operates at high temperatures that allow fuel reformation to occur, the system can internally generate hydrogen from fossil fuels.

FuelCell Energy expects to be ready to ship the fuel cell from its Torrington, Conn., fabrication plant to the Wabash River site in the second half of 2003. The project is expected to produce enough electricity to power about 2,000 homes.

DOE

Most fuel cell marketing is currently aimed at replacing batteries with fuel cell systems for back-up power applications. In off-peak hours, electricity from the power grid is used to manufacture hydrogen by electrolysis. During peak loads, or a black-out, the fuel cell kicks in to supply power. Used this way, fuel cells can deliver more power, over a longer period of time, than batteries. In other words, the energy efficiency of a fuel cell appears to be superior to the fuel efficiency of a battery for these specific applications.

Fuel cell vehicles? We have a long way to go - even if we can make it work (maybe) at a cost the average car buyer can afford (not in this decade). Toyota has already deployed several experimental vehicles. Prototype fuel cell powered vehicles are due in 2004 (Ford) and 2006 (Chrysler). But fuel cells will not be a practical power plant for vehicles until the twin problems, storage capacity and energy content, have been solved. And existing vehicle fuel cell programs use experimental technologies that may - or may not - be practical for mass deployment.

Press Release

November, 2003

"Toyota announced that two more Fuel Cell Hydrogen Vehicles (FCHV) vehicles will be put into service with the assistance of the University of California, Irvine and the University of California, Davis. The FCHVs will be leased by UC Irvine's National Fuel Cell Research Center and by UC Davis' Institute of Transportation Studies. The addition of these vehicles will bring the total number of FCHVs on the road to 18. Four will be with the universities, three with the California Fuel Cell Partnership, one at Toyota Motor Sales, U.S.A., Inc. and 10 with Japanese government agencies and private companies. Toyota's FCHVs are based on the Highlander mid-sized sport utility vehicle and contain a Toyota-developed fuel cell system with four 5,000-psi hydrogen fuel tanks. Hydrogen gas feeds into the fuel cell stack where it is combined with oxygen. This generates a peak of 90 kW of electricity that powers the 109-hp electric motor (194 lb-ft of torque) and charges the vehicle's nickel-metal hydride batteries which also feed power-on-demand to the electric motor. The only emission is clean water vapor. The FCHV has a top speed of 96 mph and a maximum range of about 180 miles."

Toyota Motor Company

It takes 10 to 15 years to launch a new technology into general consumer use. And that's after it has been completely developed. And tested. And somebody has figured out how to manufacture it. And set up a distribution system.

So we can expect that the oil crisis will be upon us long before fuel cells have emerged from the laboratory as a practical energy system.

For the intermediate term, therefore, the proven fuel efficiencies of hybrid gasoline and diesel vehicles show far greater promise as a solution to the pending oil depletion crisis. The fuel manufacturing and distribution infrastructure is already in place. Affordable and practical technology can be deployed in this decade.

Never-the-less, fuel cell research and development MUST become a national priority. For every industrialized nation. NOW.

Better late than never.

C1 Technology

The world has an abundant supply of coal. After experiencing the deprivation of the oil crisis, people are going to demand that power companies use coal for the generation of electricity.

Dirty or not.

And they will figure out a way to use it to heat their homes.

Dirty or not.

And they will use it for cooking.

Dirty or not.

So doesn't it make sense (common sense - if I dare use that word) to make sure the world's research and development resources are focused on CLEAN coal technology? I mean, if we are going to use coal anyway, why not focus on the use of clean coal technology?

We have a lot to learn.

United States, Department of Energy (DOE)

Press Release , April 29, 1999

"Research to focus on chemical processes for converting carbon materials into alternative fuels.

The U.S. Department of Energy (DOE) has awarded a $4.2 million research grant to the Consortium for Fossil Fuel Liquefaction Science for a new research program to study innovative ways to produce cleaner alternative fuels and high quality chemicals. The Consortium for Fossil Fuel Science (CFFS) has been engaged in research on the development of alternative sources for transportation fuel since 1986. The Department's Office of Fossil Energy and Office of Energy Efficiency and Renewable Energy will jointly fund the research with consortium members the University of Kentucky, West Virginia University, the University of Utah, the University of Pittsburgh, and Auburn University. These University consortium members will contribute another $1.2 million to the project.

The research will focus on C1 chemistry in the conversion of carbon-containing materials such as natural gas, coal, biomass, petroleum coke, and municipal solids wastes into useful fuels and chemicals[43]. C1 chemistry is essentially the conversion of single carbon-bearing molecules - such as those found in natural gas, carbon dioxide, or "syngas" (a mixture of carbon monoxide and hydrogen) - into fuel

---

43 Consortium for Fossil Fuel Science, 11/2003 - "C1 chemistry refers to reaction processes that use feedstock's that consist of molecules containing one carbon atom [synthesis gas (a mixture of CO and H2), methane (CH_4), carbon dioxide (CO_2), carbon monoxide (CO), and methanol (CH_3OH)].

and chemical products. For example, the process can be used to convert syngas made from coal into clean oxygenated transportation fuels that could be used in a new generation of diesel vehicles. C1 chemistry could also be used to produce high-purity hydrogen. ..."

The Consortium for Fossil Fuel Science has now focused its attention on C1 chemistry to produce ultra-clean transportation fuels and hydrogen. CFFS research has demonstrated the production of nanoscale catalysts to produce carbon nanotubes and pure hydrogen. The hydrogen could be used to power fuel cells. The carbon nanotubes show promise as a safe medium for the storage of hydrogen. The consortium has also demonstrated processes that permit fine tuning the Fischer-Tropsch (FT) reaction to produce a wider range of fuels, fuel additives, hydrogen, ethylene, propylene, and lubricants[44].

Gas To Liquids Technology

Although this technology has been around since the 1920s, it has received increasing attention over the last five years because it offers a way to increase the supply of natural gas and convert other hydrocarbons to vehicle fuel. It is technically possible, but not necessarily practical, to synthesis almost any hydrocarbon from any other hydrocarbon. Natural gas can be converted to a liquid hydrocarbon. Coal can be converted to diesel fuel. A methane hydrocarbon (such as natural gas at the well head) may be directly converted to a synthetic petroleum product (syncrude), or indirectly through the Fischer-Tropsch (FT) reaction to synthesis gas (syngas) [45]. In the past, both processes were expensive, capital intensive and difficult to control.

It has often been pointed out we have an abundance of natural gas. Unfortunately, much of this natural gas is located in regions where there are no pipelines to transport it to consumers, or it may be located in sources that are too small to be commercially attractive. Qatar, for example, has huge deposits of natural gas. In order to transport this gas to markets in Western Europe, North America and the Asia/Pacific region, it must be liquefied. The conventional way was to liquefy the natural gas (LNG) in pressurized containers. With the GTL process, it can be shipped more efficiently as either a feedstock or as a diesel fuel.

During the refining of crude oil, associated-dissolved, or AD, gas is produced as a byproduct. The easiest way to dispose of this gas is to

---

44 Consortium for Fossil Fuel Science, University of Kentucky, 533 South Limestone Street, Lexington, KY 40508-4005. http://www.cffls.uky.edu/

45 For more information: http://www.fischer-tropsch.org, http://www.ivanhoe-energy.com, http://www.sasolchevrn.com.

vent it or flare (burn) it, releasing methane and carbon dioxide into the air. However, a GTL facility can use this gas as a feedstock, thus reducing the need for flaring or venting.

Conventional oil based diesel fuel emissions contain particulate matter and aromatics. GTL technology, by contrast, is appealing because it would produce hydrocarbon chains that are free from sulfur, metals and aromatics. With its high cetane number and inherent purity, GTL diesel would command a premium as a liquid that can be blended with conventional oil based diesel fuel to reduce particulate and aromatic emissions. As a result, during combustion it produces less particulate matter and aromatics. GTL diesel can be combined with conventional diesel fuel for use in existing diesel engines, or - with its higher cetane - used in a higher performance diesel engine.

Synthetic fuels can be substituted for methanol in fuel cells. GTL naphtha functions as a good petrochemical feedstock. The waxy portion of the resulting production can be converted to high value products such as lubricants, drilling fluids, and waxes.

There is a growing interest in the conversion of coal to diesel fuel. Coal is an important source of synthesis gas, a mixture of hydrogen and carbon monoxide. It reacts in the presence of an iron or cobalt catalyst, producing heat (which can be used to generate electric power), and methane, synthetic gasoline, synthetic diesel fuel and alcohols. Water and carbon dioxide are released as a by-product.

So if there are so many benefits, why haven't we seen more excitement about GTL technology? GTL diesel produced from natural gas has a projected cost of less than $25 per barrel. GTL diesel produced from coal, which involves an extra process, would have a cost of around $35 per barrel. Neither were competitive in price with conventional oil based products.

But that is changing. Because of its purity, GTL diesel will command a premium wherever local environmental legislation has set a cap on emissions from diesel vehicles. If the goal is to reduce sulfur from 500 parts per million to under 100 parts per million, it is doubtful these improvements can be made without some blending of zero sulfur GTL diesel with conventional oil based diesel.

Chevron (ChevronTexaco) believes gas-to-liquids technology is so promising that its development could create a paradigm shift within the petroleum industry. It could be used to commercialize the trillions of cubic feet of natural gas throughout the world that are currently isolated from the traditional gas infrastructure. Natural gas can also be converted into superior fuels, including no-sulfur diesel.

Through a joint venture with Sasol, Chevron intends to create additional options for dealing with existing reserves of natural gas, the recovery of gas that is being flared as a by-product of petroleum refining operations, and the development of natural gas reserves where there isn't any gas distribution infrastructure. The GTL technology converts natural gas to synthesis gas, then converts synthesis gas to a waxy synthetic crude, using Sasol's Fischer-Tropsch technology, and then upgrades the synthetic crude to high-quality diesel and naphtha, using Chevron's hydroprocessing technology.

Several other oil company efforts are under way. For example:

U.S. Department of Energy Global Forum on Personal Transportation
Remarks for Shell Oil Company, President Rob Routs, November 12, 2002

... "In the medium term, the recent advances with Fischer-Tropsch Gas-to-Liquids (GTL) technology could bring a new perspective to personal transportation. GTL ultra-clean diesel is not just a cleaner version of refinery diesel; it is a water-white, gas-based fuel, virtually free of sulfur and aromatics content. This gas-based fuel can be used in existing diesel engines and distribution infrastructure. It is considered the most cost-effective alternate fuel for the foreseeable future.

Studies show that GTL ultra-clean diesel, when blended to standard diesel in automotive applications, can significantly reduce the emissions of particle matters, carbon monoxides and hydrocarbon. A recent Shell-sponsored comprehensive life-cycle assessment by PricewaterhouseCoopers compared a GTL system to a crude oil refinery system. The GTL system has less impact on air acidification, far lower impact on smog formation and no greater impact on global warming.

GTL fuels are compatible with many possible directions of the transportation fuel market, since they can be used in standard diesel engines, either as a blend with standard diesel fuel or as a 100 percent product; in advanced engine design including diesel/electric hybrids and hydrocarbon-powered fuel cells. GTL plants also can produce a range of specialty products, such as normal paraffins and lubricant base oils.

Shell is a significant player in GTL technology. .... Shell's state-of-the-art proprietary GTL process - the Shell Middle Distillate Synthesis (SMDS) - has been used at our commercial-scale plant in Malaysia. This is the world's first medium-scale Fischer-Tropsch GTL plant. Recent breakthroughs have significantly reduced the unit capex of GTL, and we now are pursuing construction of second generation world-scale SMDS plants." ...

Fischer-Tropsch Gas-to-Liquids (GTL) technology could provide fuels for public transportation, the use of personal vehicles, and fixed site engine applications. Abundant coal resources offer an attractive mid-term option for producing the large quantities of hydrogen that will be required. Initially, hydrogen will be produced in coal gasification facilities capable of co-producing electric power and other high-value fuels and chemicals. A key element of the hydrogen initiative is to develop advanced hydrogen production and delivery technologies that can supply energy and transportation systems with affordable hydrogen at significantly reduced emissions. With carbon sequestration, it may be possible to use coal as a feedstock in the production of hydrogen for many decades without adding large volumes of carbon gases to the atmosphere.

The cost of coal gasification has declined from $60.00 per barrel of oil equivalent using 1970s technology, to $40.00 per barrel using 2003 technology. The DOE's goal is to get the cost down below $30.00 per barrel of oil equivalent by 2010.

Appendix 3. Definitions

Cultural Economics

The study of how human culture interacts with economic events and conditions. Culture, in this sense, includes everything we are: our political systems, religious beliefs, ethnic character, mores, traditions, history, customs, arts, sciences, and education. These all play a role in how we chose to organize the production of goods and services, the values we place on labor and opportunity, how we make purchase and investment decisions, and how we utilize the resources of this earth. The term "Economics" refers to the extent and process of how we employ capital, labor and materials. In the aggregate, these drive the data that is used to measure how our economy is behaving - markets, raw materials, production, finished goods, revenues, costs, profits, inventory, employment, housing, income, savings, stocks, bonds - and so on.

Geography

For this the purposes of this report, there are four oil producer regions.

- North America - includes oil produced by the U. S. A., Canada and Mexico

- Middle East - includes oil produced by the United Arab Emirates, Bahrain, Tunisia, Algeria, Saudi Arabia, Syria, Iraq, Qatar, Kuwait, Libya and Egypt - sometimes known as OAPEC - as well as Jordan, Sudan, Oman, Morocco and Yemen.

- EurAsia - includes oil produced by Eastern and Western Europe and all of the countries included in the former Soviet Union.

- ROW - Rest of World - includes oil produced by the nations that border the Western Pacific ocean (China, India, Australia, S. Korea, Vietnam, Japan, Indonesia, etc.), Central and South America, and Africa (not included in the Middle East).

There are four oil consumer regions.

- North America - includes oil consumed by the U. S. A.,
  Canada and Mexico

- Western Europe- For the purposes of this report, this area
  includes members of the European Union (15 countries)
  and nations that are members of the European Free Trade
  Association (4 countries).

- Asia/Pacific- Nations that abut the western edge of the
  Pacific Rim, including Taiwan, S. Korea, Singapore,
  Malaysia, Indonesia, China, India, Japan and Australia

- ROW- any nation not found in the above three geographic
  areas including most of the former Soviet Union, Central
  and South America, the Middle East and Africa.

Industry Definitions

Reserves
(From the Society of Petroleum Engineers Website spe.org)

"Reserves are those quantities of petroleum which are anticipated to be
commercially recovered from known accumulations from a given date
forward. All reserve estimates involve some degree of uncertainty. The
uncertainty depends chiefly on the amount of reliable geologic and
engineering data available at the time of the estimate and the
interpretation of these data. The relative degree of uncertainty may be
conveyed by placing reserves into one of two principal classifications,
either proved or unproved. Unproved reserves are less certain to be
recovered than proved reserves and may be further sub-classified as
probable and possible reserves to denote progressively increasing
uncertainty in their recoverability. ...

Estimation of reserves is done under conditions of uncertainty. The
method of estimation is called deterministic if a single best estimate of
reserves is made based on known geological, engineering, and
economic data. The method of estimation is called probabilistic when
the known geological, engineering, and economic data are used to
generate a range of estimates and their associated probabilities.
Identifying reserves as proved, probable, and possible has been the

most frequent classification method and gives an indication of the probability of recovery. Because of potential differences in uncertainty, caution should be exercised when aggregating reserves of different classifications. ...

Proved Reserves

Proved reserves are those quantities of petroleum which, by analysis of geological and engineering data, can be estimated with reasonable (90%) certainty to be commercially recoverable, from a given date forward, from known reservoirs and under current economic conditions, operating methods, and government regulations. Proved reserves can be categorized as developed or undeveloped. ....

Unproved Reserves

Unproved reserves are based on geologic and/or engineering data similar to that used in estimates of proved reserves; but technical, contractual, economic, or regulatory uncertainties preclude such reserves being classified as proved. Unproved reserves may be further classified as probable reserves and possible reserves. ...

Probable Reserves

Probable reserves are those unproved reserves which analysis of geological and engineering data suggests are more likely than not to be recoverable. In this context, when probabilistic methods are used, there should be at least a 50% probability that the quantities actually recovered will equal or exceed the sum of estimated proved plus probable reserves. ...

Possible Reserves

Possible reserves are those unproved reserves which analysis of geological and engineering data suggests are less likely to be recoverable than probable reserves. In this context, when probabilistic methods are used, there should be at least a 10% probability that the quantities actually recovered will equal or exceed the sum of estimated proved plus probable plus possible reserves. ...""

Approved by the Board of Directors, Society of Petroleum Engineers (SPE) Inc., and the Executive Board, World Petroleum Congresses (WPC), March 1997

Terms

Many of the definitions below are provided courtesy of the American Petroleum Institute http://api.org, and are derived from Introduction to Oil and Gas Production, Book One of the Vocational Training Series, Fifth Edition, June 1996.

Barrel – a unit of measure for oil and petroleum products that is equivalent to 42 U.S. gallons.

Bbl - *Billion barrels*

Development well – a well drilled within the proved area of an oil or gas reservoir to the depth of a stratigraphic horizon known to be productive; a well drilled in a proven field for the purpose of completing the desired spacing pattern of production.

Downstream – when referring to the oil and gas industry, this term indicates the refining and marketing sectors of the industry. More generically, the term can be used to refer to any step further along in the process.

Dry hole – any exploratory or development well that does not find commercial quantities of hydrocarbons.

E&P - Exploration and production. The "upstream" sector of the oil and gas industry.

Enhanced oil recovery (EOR) – refers to a variety of processes to increase the amount of oil removed from a reservoir, typically by injecting a liquid (e.g., water, surfactant) or gas (e.g., nitrogen, carbon dioxide).

Exploratory well – a hole drilled: a) to find and produce oil or gas in an area previously considered unproductive area; b) to find a new reservoir in a known field, i.e., one previously producing oil and gas from another reservoir, or c) to extend the limit of a known oil or gas reservoir.

Field – An area consisting of a single reservoir or multiple reservoirs all grouped on, or related to, the same individual geological structural feature or stratigraphic condition. The field name refers to the surface area, although it may refer to both the surface and the underground productive formations.

Mbl - *Million barrels*

Mbpd - *Million barrels per day, a measure of the volume of oil pumped per day.*

Natural gas liquids (NGL) – the portions of gas from a reservoir that are liquefied at the surface in separators, field facilities, or gas processing plants. NGL from gas processing plants is also called liquefied petroleum gas (LPG).

OAPEC - *Organization of Arab Petroleum Exporting Countries*

OPEC - Organization of Petroleum Exporting Countries

OIL SHALE -- A type of rock containing organic matter that produces large amounts of oil when heated to high temperatures.

Sour crude oil – oil containing free sulfur or other sulfur compounds whose total sulfur content is in excess of 1 percent.

Tbl - Trillion barrels

Viscosity - a measure of a fluid's resistance to flow. It describes the internal friction of a moving fluid. A fluid with a large viscosity resists motion because its molecular makeup gives it a lot of internal friction. A fluid with low viscosity flows easily because its molecular makeup results in very little friction when it is in motion. The oil industry measures viscosity against a standard established by the American Petroleum Institute, called the API. Included in the calculation is oil specific gravity and oil density. Water has an API of 10. If a liquid is less dense than water, then the API > 10. If the liquid is denser than water, then the API < 10. Typical Oil API Gravities: Heavy oil < 20 API (dense), Black oil < 40 API (very dense), Volatile oil > 40 API (flows like a gas), and Condensates ~ 40 - 60 API. Oils with API Gravities in the range > 35 - 45 and free of contaminants are highly desired.

Wildcat well – a well drilled in an area where no current oil or gas production exists. Also called a "rank wildcat."

References

Material for Chapters 1 and 5 was researched from the following references:

The Heritage Foundation, http://www.heritage.org/Research/MiddleEast/wm217.cfm, includes a good article and a long list of Iraq trade references.

mensnewsdaily.com www.mensnewsdaily.com Article on Saudi Arabia by Paul C. Robbins Ph. D., May 7, 2003

Price of Honor, Muslim Women Lift the Veil of Silence on the Islamic World, by Jan Goodwin, Little Brown and Company, 1994.

See No Evil: The True Story of a Ground Soldier in the CIA's War on Terrorism, Robert Baer, Crown Publishers.

Sleeping With The Devil, How Washington Sold Our Soul for Saudi Crude, Robert Baer, Crown Publishers.

The fall of the House of Saud, Robert Baer, The Atlantic Monthly, May 2003, and Sleeping With The Devil, How Washington Sold Our Soul for Saudi Crude, Robert Baer, Crown Publishers.

Stockholm International Peace Research Institute (SIPRI), www.sipri.se, "Arms Transfers to Iraq, 1981–2001," at http://projects.sipri.se/armstrade/IRQ_IMPORTS_1982-2001.pdf.

The National Review, www.nationalreview.com, Articles on Saudi Arabia and terrorism.

Arab Human Development Report 2003, UNDP, United Nations Development Program, www.undp.org, is a thoughtful analysis of cultural challenges facing Arab nations.

Central Intelligence Agency, The World Factbook 2002, http//www.cia.gov/cia/publications/factbook.

Kenneth Katzman, Iraq: Oil-for-Food Program, International Sanctions, and Illicit Trade, Congressional Research Service, September 26, 2002.

Faye Bowers, "Driving Forces in War-wary Nations: The Stances of France, Germany, Russia and China are colored by economic and national interests", Christian Science Monitor, February 25, 2003.

Dan Morgan and David B. Ottaway, "In Iraqi War Scenario, Oil Is Key Issue," The Washington Post, September 15, 2002.

Partial list of additional references:

A History of the Arab Peoples
Albert Hourani, The Belknap Press of Harvard University Press, 1991

Alexander's Oil and Gas Connections
www.gasandoil.com
World reserves and production. Oil industry activity and issues.

American Petroleum Institute
www.api.org
Information about oil and gas for public and industry professional use.

American Solar Energy Society
http://www.ases.org/
The American Solar Energy Society (ASES) promotes the widespread near-term and long-term use of solar energy. ASES is the United States Section of the International Solar Energy Society.

ASPO, The Association for the Study of Peak Oil & Gas
www.peakoil.net
Conference, news and reports on oil and gas depletion.

BP
www.bp.com
Publishes historical data on oil production and consumption as well as oil refining and distribution issues.

Business Week
www.businessweek.com
Articles and commentary on business and current events.

Cambridge Energy Research Associates
www.cera.com
Cambridge Energy Research Associates (CERA) is an international consulting firm that focuses on energy markets, geopolitics, structure, and strategy.

CGES
Centre for Global Energy Studies
www.cges.co.uk
Provides independent and objective information and analysis on
the key energy issues.

Congressional Research Service, The Library of Congress
The Congressional Research Service is the public policy research
arm of the United States Congress. As a legislative branch agency
within the Library of Congress, CRS works exclusively and directly for
Members of Congress, their Committees and staff on a confidential,
nonpartisan basis.

Consortium for Fossil Fuel Science
http://www.cffls.uky.edu/
Research on the development of alternative sources for
transportation fuel, including coal, natural gas, and waste materials.

Douglas-Westwood
www.dw-1.com
World Oil Report
Oil consumption and production forecast, oil industry issues and
information.

Economy.com, Inc.,
www.economy.com
Economic data and research.

EconStats
http://www.econstats.com/
Economic Data

EnCana Corporation
www.encana.com
A leader in the development of Steam Assisted Gravity Drainage
(SAGD) for the production of oil from tar sands.

Energy Information Administration
www.eia.doe.gov
Production, consumption and price data

Energy Security Analysis Inc.

www.esai.com

Monitors, analyzes and synthesizes information about worldwide energy markets.

EVWorld.com,Inc.,

editor@evworld.com.

EV World™ and EV World Update include running commentaries on energy subjects.

From the Wilderness, Mike Ruppert

www.fromthewilderness.com

www.copvcia.com

WEB sites of an investigative journalist. He publishes a monthly newsletter.

Glenn R. Morton, author: The Coming Energy Crisis, From Perspectives on Science and Christian Faith, 52 (December 2000): 228-229. Articles and commentary by Glenn Morton can be found at asa3.org/archive/asa/200007/0167.html

Global Insight

http://www.globalinsight.com/

Forecasts, analysis and consulting to business and governments around the world.

IHS Energy

www.ihsenergy.com

Worldwide Exploration & Production databases and decision-support solutions.

International Association for Energy Economics

http://www.iaee.org/

Provides a forum for the exchange of ideas, experience and issues among professionals interested in energy economics.

International Energy Agency

www.iea.org

Based in Paris, IEA is an autonomous agency linked with the Organization for Economic Co-operation and Development (OECD). The IEA produces the annual World Energy Outlook.

National Council for Science and the Environment
www.ncseonline.org
The Council envisions a society where environmental decisions are based on an accurate understanding of the underlying science, its meaning, and its limitations.

National Energy Technology Laboratory
www.netl.doe.gov
The development of clean fuel technologies related to coal and natural gas.

National Hydrogen Association
http://www.hydrogenus.com/
The National Hydrogen Association, works to create a shared vision for a hydrogen energy infrastructure.

NREL
http://www.nrel.gov/
The National Renewable Energy Laboratory (NREL) is a leader in the U.S. Department of Energy's effort to secure an energy future for the nation that is environmentally and economically sustainable.

Office of Fossil Energy (FE), U.S. Department of Energy
www.fe.doe.gov
Responsible for the development of clean coal technologies, a pollution-free plant to co-produce electricity and hydrogen, and the nation's Strategic Petroleum reserve.

Oil & Gas Journal, is a weekly magazine of international petroleum news and technology, published by PennWell Corporation, 1700 West Loop South, Suite 1000, Houston TX 77027,
http://ogj.pennnet.com/home.cfm

Saudi-American Forum
http://www.saudi-american-forum.org
An information service that provides information on the U.S.-Saudi relationship.

SEPA

Solar Electric Power Association

SEPA is a collaboration of utility, energy service, and photovoltaic industry participants that encourages the commercial use of solar electric power.

Simmons & Company International

http://www.simmonsco-intl.com

Energy industry investment banking services. The world's largest Energy Investment Bank.

Stockholm International Peace Research Institute (SIPRI), www.sipri.se

Conducts research on questions of conflict and cooperation of importance for international peace.

The National Review

WEB Site   www.nationalreview.com

Articles on Saudi Arabia and terrorism.

U.S. Department Of Labor

Bureau of Labor Statistics

www.bls.gov

Unemployment and Consumer Price Index data.

Upstream

www.upstream.tm

An independent news service for the oil and gas industry.

Why I Am Not A Muslim, by Ibn Warraq, Prometheus Books, 1995

World Bank

http://www.worldbank.org/data/

Publishes a range of indicators relating to development, including economic and social statistics.

World Oil, PO Box 2608, Houston, Texas 77252, http://www.worldoil.com/, and World Oil Magazine, published by Gulf Publishing Company, provides monthly editorial content and handbooks on petroleum exploration, drilling and production.

World Energy Council
http://www.worldenergy.org/
Studies and conferences on energy.

World Resources Institute
www.wri.org
An environmental policy organization that promotes solutions to protect the Earth.

WTRG Economics
www.wtrg.com
Analysis, planning, forecast and data services for energy producers and consumers.